Ethical Governance of Artificial Intelligence in the Public Sector

This book argues that ethical evaluation of AI should be an integral part of public service ethics and that an effective normative framework is needed to provide ethical principles and evaluation for decision-making in the public sphere, at both local and international levels.

It introduces how the tenets of prudential rationality ethics, through critical engagement with intersectionality, can contribute to a more successful negotiation of the challenges created by technological innovations in AI and afford a relational, interactive, flexible and fluid framework that meets the features of AI research projects, so that core public and individual values are still honoured in the face of technological development.

This book will be of key interest to scholars, students, and professionals engaged in public management and ethics management, AI ethics, public organizations, public service leadership and more broadly to public administration and policy, as well as applied ethics and philosophy.

Liza Ireni-Saban is the deputy dean and a senior lecturer at the Lauder School of Government, Diplomacy and Strategy, Interdisciplinary Centre (IDC) in Herzliya, Israel.

Maya Sherman is a student in Government, Diplomacy and Strategy at the Interdisciplinary Centre (IDC) in Herzliya, Israel.

Routledge Studies in Policy and Power

This series presents state-of-the-art analyses of the relationship between policy, politics and power. Transcending disciplinary boundaries, it recognises that policy formation is fundamentally a product of contestation between different social forces, interests, ideas and ideologies. Moreover, the implementation of policy is conditioned by these same elements, yielding actual policy outcomes and patterns of behavioral change that often deviate from both intentions and theoretical assumptions. The series encourages critically-oriented submissions focused on the latest developments in policymaking, implementation and outcomes from around the world.

Series editors: *Toby Carroll, City University of Hong Kong, Hong Kong, Kelly Gerard, University of Western Australia, Australia, and Darryl S.L. Jarvis, Hamad Bin Khalifa University (HBKU), Qatar Foundation, Doha.*

Ethical Governance of Artificial Intelligence in the Public Sector
Liza Ireni-Saban and Maya Sherman

Ethical Governance of Artificial Intelligence in the Public Sector

Liza Ireni-Saban and Maya Sherman

Routledge
Taylor & Francis Group

LONDON AND NEW YORK

First published 2022
by Routledge
2 Park Square, Milton Park, Abingdon, Oxon OX14 4RN

and by Routledge
605 Third Avenue, New York, NY 10158

Routledge is an imprint of the Taylor & Francis Group, an informa business

© 2022 Liza Ireni-Saban and Maya Sherman

The right of Liza Ireni-Saban and Maya Sherman to be identified as authors of this work has been asserted by them in accordance with sections 77 and 78 of the Copyright, Designs and Patents Act 1988.

British Library Cataloguing-in-Publication Data
A catalogue record for this book is available from the British Library

Library of Congress Cataloging-in-Publication Data
Names: Ireni Saban, Liza, author. | Sherman, Maya, author.
Title: Ethical governance of artificial intelligence in the
public sector / Liza Ireni Saban and Maya Sherman.
Description: Milton Park, Abingdon, Oxon ; New York, NY : Routledge, 2022. |
Series: Routledge studies in policy and power |
Includes bibliographical references and index.
Identifiers: LCCN 2021013604 (print) | LCCN 2021013605 (ebook) |
ISBN 9780367618087 (hardback) | ISBN 9780367618131 (paperback) |
ISBN 9781003106678 (ebook)
Subjects: LCSH: Artificial intelligence–Moral and ethical
aspects. | Artificial intelligence–Government policy.
Classification: LCC Q334.7 .I74 2022 (print) |
LCC Q334.7 (ebook) | DDC 172/.2–dc23
LC record available at https://lccn.loc.gov/2021013604
LC ebook record available at https://lccn.loc.gov/2021013605

ISBN: 978-0-367-61808-7 (hbk)
ISBN: 978-0-367-61813-1 (pbk)
ISBN: 978-1-003-10667-8 (ebk)

DOI: 10.4324/9781003106678

Typeset in Times New Roman
by Newgen Publishing UK

Contents

Introduction

The growth of Artificial Intelligence (AI) applications has led to unprecedented social, ethical and legal implications, as recently seen in salient technological affairs in the public sphere. Controversial affairs such as the Cambridge Analytica data scandal and the autonomous car fatality in Arizona have brought the ethical dimensions of advanced technological developments to the forefront. Experts both from within the tech world and without having started to confront issues that drive home the anxiety society faces with unfettered technological progress: What are the ethical boundaries of technological determinism? and, What philosophical theory can correctly channel human ethical behaviour against the backdrop of exponential technological development?

The scholarly discourse has not properly addressed the impact of AI's technological loopholes on the public sector and the needed governmental and civil confrontation. AI technologies are often criticized due to their potentially biased and harmful implications, which impact certain groups of individuals. A substantial body of research has elaborated upon the inherent loopholes of these technologies, with emphasis on AI bias, transparency and accountability. Important to note, AI does not necessarily create new bias, but may reinforce existing human bias, since humans are relatively autonomous and flawed creatures employing these technologies to their own ends.

Due to the growing use of AI in the public sector, these challenges have the potential to undermine the appropriateness of decision-making processes in the public sphere, and therefore must be properly tackled or moderated at the local and international levels. An effective normative framework must provide ethical principles of decision-making, suggest applications appropriate for use, and acknowledge the dignity, rights and public interest at stake in the situated use of AI techniques. Therefore, it is of paramount importance to set an ethical framework that is flexible enough to accommodate these challenges and ensure

DOI: 10.4324/9781003106678-1

that core public and individual values are still honored in the face of technological development, in order to provide practical solutions to the binary and government implications parallelly.

The question is then, how can tools of ethical governance in contemporary democracies bridge the gap between the tech giants, whose initial priority is to monetize their products and platforms, and the public good, highlighting the civil will and one's inherent rights? Compared to alternative and more traditional ethical approaches, we suggest that prudential rationality ethics can become a means for public administrators to negotiate the challenges posed by the implementation of AI technologies in the public sector. The prudential aspect of rationality brings a relational, interactive, flexible and fluid framework that meets the features of AI research projects. We propose that the ethics of prudential rationality could develop familiarity with the theoretical lens of intersectionality. Our aim in this book is to explore how prudential ethics can be enhanced through critical engagement with intersectionality in a way that the biases and problems built into AI techniques are clearly evaluated and framed in the public service.

In this book, we aim to show that prudential ethics is inspired by intersectionality; both approaches share the normative grounding of social justice. Specifically, we aim to introduce how the normative tenets of prudential rationality ethics can benefit from intersectionality how to move beyond binary thinking in AI and address multiple grounds of diversity and inequality.

Specifically, we suggest that prudential rationality furnishes the most comprehensive and appropriate approach for dealing with the arising ethical issues which have been, and will continue to be, emerge from unrestricted technological growth and capability. Prudential rationality, we go on to suggest, is far more advantageous than the other ethical models of utilitarianism and deontology in confronting the challenges of the impending AI era.

Indeed, humankind has come to accept the reality that ever-advancing technology is a fixture of modernized societies. As Merritt Roe Smith and Leo Marx (1994) point out, although people may take for granted the omnipresence and perma-developing nature of technology as an inherent part of the modern human experience, it, in fact, represents a unique notion—'technological determinism'. The power of technology as an agent for change has been crystallized in humanity's consciousness and plays a central role in the culture of modernity. The advent of the computer in the 20th century was a formative occurrence which reshaped and renegotiated the trajectory of the human experience and surely defined that century and those to come. The tech revolution affected every aspect of the human experience in a relatively short period of time. Computers have changed the way in which mankind communicates, interacts and experiences knowledge. Even the conception of information and how it is transmitted has undergone a paradigm shift as a result of the computer era. Whereas once the transmission and consumption of knowledge were restricted to academic settings, which often allowed limited access due to economic or social restrictions, computers have enabled virtually unfettered access to unlimited amounts of information. However, as paramount and inseparable as the computer revolution seems to the experience of modernity, the notion of technology as an independent entity and a 'virtually autonomous agent of change' is a phenomenon with roots that reach far back into human history (Smith & Marx, 1994: 11).

We are not unfamiliar with standing tropes and popular narratives which conceive of technology as a standalone force which guides history. The human experience has been peppered with technological advancements that have fundamentally shifted the way humans obtain knowledge and altered the course of human history. The Mechanical Revolution ushered in the era of discovery and exploration, the printing press is seen as the cause of the Reformation, Whitney's cotton gin is causally linked to the American Civil War and the advent of electricity is conceived of as the lifeblood of the Second Industrial Revolution and the hallmark of the industrialized Western world. Technological determinism, as such, is the perception that technology is 'a driving force of history', where the narrative emphasis is put on technology as, 'the material artifact and the changes it presumably effects' (Smith & Marx, 1994: 10). Merritt Roe Smith posits that the corporality of technology lends to the palpable sense of technological developments as discrete catalysts for change in points of history. The absoluteness of technology's materiality ascribes to it a more determinative character than other, more abstract catalysts for change, such as ideological, political or social currents. Because of the nature of technological determinism, each development is perceived as having facilitated the next. As a result, the dependence of society both conceptually and infrastructurally has become increasingly more intertwined with interconnected technological systems. Technological dependence has become so fundamental and ubiquitous that, as Smith puts forth, '[t]he megasystem's continued functioning is a precondition for the reproduction of the entire social order' (Smith & Marx, 1994: 12).

Whether technological development, a defined driving force in history is a heuristic, as Heilbroner (1994) puts it, or not, there is no question that modern technological development has widened the conceptual and concrete boundaries of society. As is often the case, technological breakthroughs precede the systems and mechanisms for dealing with the new realities and consequences which they create. This trend is demonstrated perfectly by the development of the first nuclear weapons. Major General Leslie Groves and Robert Oppenheimer's Manhattan Project was framed as 'the weapon which ended the War and…raised the curtain on vistas of the new world' (Hewlett & Anderson, 1962). However, as Joseph Masco (2013) pointedly notes in his book *The Nuclear Borderlands: The Manhattan Project in Post-Cold War New Mexico*, the perception of the invention of the A-bomb quickly shifted from technological glorification cushioned in the prose of 'national security' and 'mutually assured destruction' to 'unthinkable' after Hiroshima and Nagasaki. The development of new technologies, no matter how well-intentioned their predicted application may be, inevitably ushers in a seemingly permanently reactive response for unforeseen consequences. This begs the question: Must we always suffer a great loss in order to correctly deal with humanity's evolving reality?

Despite our best efforts to create systems and structures for dealing with the new realities created by the constantly evolving computer era, we are still lagging. This has become glaringly clear in light of recent tech scandals which have had implications across almost every segment of society. The Cambridge Analytica scandal, the most conspicuous involving the consequences of unfettered and unregulated tech, is hardly the first, nor the last. However, in the popular consciousness, it is the watershed moment for developing the understanding of the potential abuses that can occur due to our dependence on technological megasystems and our unpreparedness for consequences. The Cambridge Analytica scandal broke onto the global stage in 2018 when a whistleblower from within the company revealed that it had harvested data from millions of Facebook profiles for political means without users' consent. Although *Guardian* reporter Harry Davies had revealed that Cambridge Analytica (CA) was engaging in this kind of illegal behaviour as early as 2015, it only received mass international attention and outrage following the revelations from the whistleblower and former CA employee, Christopher Wylie (Cadwalladr, 2017; Davies, 2015).

The original 2015 article delineated how US politician Ted Cruz was employing a firm that used harvested psychological data to urge tens of millions of voters to vote for Cruz. The article details that Cruz's firm, backed by a billionaire benefactor, paid academics at Cambridge University to 'gather detailed psychological profiles about the US electorate using a massive pool of mainly unwitting US Facebook users built with an online survey' (Davies, 2015). However, the scandal exploded onto the international stage only in early 2018, when several media purveyors broke the story simultaneously on March 17, 2018 (Rosenberg, Confessore, &Cadwalladr, 2018).

The article detailed that the relatively unknown company SCL had been bought out by secretive hedge fund billionaire Robert Mercer in 2013 and was promptly renamed Cambridge Analytica (ibid.). The article revealed that this big data analytics company was responsible for both the Trump and Brexit campaigns. While in our collective consciousness, the CA scandal may be conceived of as massive data theft, Cadwalladr rightly predicted that it had more far-reaching consequences which she separated into three discrete points:

> *How the foundations of an authoritarian surveillance state are being laid in the US. How British democracy was subverted through a covert plan of coordination enabled by a US billionaire. And how we are in the midst of a massive land grab for power by billionaires via our data…which is being silently amassed, harvested and stored.*
>
> (ibid., 9)

Cadwalladr's article did an excellent job of putting the CA ordeal into perspective as something far more sinister than simple campaign research; CA had captured and exploited individuals' data and it had played an integral role in the Trump and Brexit victories (Doward, Cadwalladr, & Gibbs, 2017). Elon University Professor of Communications Jonathan Albright referred to this company as the political right's 'propaganda machine'. Cadwalladr bolstered this suggestion with her assertion that Mercer and Steve Bannon were strategically and methodically using CA to quash mainstream media and subvert it with right-wing propaganda. Cadwalladr's May 2017 article prompted two investigations into the company's behaviour, one by the Information Commissioner's Office and the second by the Electoral Commission. The former came on the heels of complaints from privacy activists, politicians and legal experts who feared that CA's role in the Brexit vote was indicative of the lag of electoral laws behind the technology. The article detailed how the communications director of Leave.EU, the pro-Brexit group, had been introduced to Cambridge Analytica by the Mercer family (ibid., 13). On the company's part, Cambridge Analytica's CEO Alexander Nix commented in 2016 that their campaign for Leave.EU focused on 'ensuring the right messages are getting to the right voters online' and remarked about targeting specific users, 'the more you know about your target audience, the better you will be able to engage, persuade and motivate them to act' (ibid., 16).

So how did a seemingly innocuous strategy for audience targeting turn into the dystopian panoptic data monster that shocked the world in 2018? The *New York Times* article which co-broke the story revealed that CA had, 'harvested private information from the Facebook profiles of more than 50 million users without their permission' (Rosenberg, Confessore, & Cadwalladr, 2018). The data was amassed when users took a personality quiz app on Facebook designed by Aleksander Kogan, the Cambridge University professor who had started SCL (Cadwalladr & Graham-Harrison, 2018). Kogan used a portion of the information gleaned from this app to create psychological profiles for CA to target US voters with political messages (Lomas, 2018). Not only was this described as one of the biggest data leaks in Facebook's history, but it also afforded Cambridge Analytica the information it needed to develop hauntingly accurate profiles for targeting the American electorate. The article also revealed that the firm had received USD 15 million from Robert Mercer in order to create a political weapon that could— and did—allow wealthy investors the ability to reshape politics. Wylie, who leaked the initial story to Cadwalladr, said of CA's management, 'Rules don't matter for them. For them. This is a war, and it's all fair … Cambridge Analytica was supposed to be the arsenal of weapons to fight that culture war'. An article published in 2017 by *The Intercept* broke down CA's system for classifying users based on their data. The system, referred to as OCEAN, classified users using the personality traits of openness, conscientiousness, extroversion, agreeableness and neuroticism. (Schwartz, 2017) Although the efficacy of the OCEAN method has been questioned, CA boasted that its results could be used to inform 'microtargeting' which can drive voter behaviour. Moreover, Nix revealed that 'neurotic voters tend to be moved by "rational and fear-based" arguments, while introverted, agreeable voters are more susceptible to "tradition and habits and family and community"'' (ibid., 22).

Cambridge Analytica was not the only company to suffer widespread public blowback. The news articles also brought on a firestorm of back-lash for Facebook for failing to protect users' data. *The Intercept* article suggested that Facebook was tacitly participated in the breach of more than 30 million users by letting producers take advantage of its porous data usage permissions for developers. After the 2015 *Guardian* article which revealed that SCL had harvested data from the profiles of unaware Facebook users, Facebook contacted SCL and requested that it delete all the data it had harvested for the Ted Cruz campaign; however, it is unclear whether SCL bided their request. Publicly, Facebook maintained that no wrongdoing had been committed by SCL when it was 'carefully investigating this situation' in 2015 (ibid., 21). However, Facebook did little to restrain third party access to its users' data, nor, according to *Tech Crunch*, did it offer more transparency into its strategy for how its platform was being abused. Mark Zuckerberg even slammed the idea that his social media platform was being used to proliferate 'fake news' as 'a pretty crazy idea' (ibid., 21). However, this was at the same time that Facebook supplied the 2016 Trump campaign offices with Facebook employees to help its digital team harness Facebook ads to their benefit (Beckett, 2017; Deagon, 2018). Immediately after the March 2018 stories broke, Facebook announced that it was suspending CA's access to its platform. However, it also sternly requested that *The Guardian* not refer to the incident as a 'breach'.

Although Facebook had issued a stream of statements promising to tighten control of its privacy regulations over the last decade, powerful players and politicians began to demand an answer from Zuckerberg as to why his platform could not product users, particularly if it was complicit in helping Trump run on a political platform of a divisive culture war riddled with populism and racism. Instead of answers, Facebook's executives remained deafeningly silent in the immediate aftermath of the CA story. Yet public opinion of the company was resoundingly loud as its market value dropped USD 36 billion over a three-day period (ibid., 28). On Tuesday, April 10, 2018, Zuckerberg appeared before the US Congress. The CEO answered policy-makers' questions regarding the breach, his company's policies on privacy, data mining, Cambridge Analytica and third party access in a five-hour testimony. Over the course of the hearing, Zuckerberg took responsibility for the CA scandal and admitted that it was a mistake not to ban the company's access to the platform (Watson, 2018).

The outcry against violations of privacy following the Cambridge Analytica Facebook data breach has been resounding. The political elite and grass-roots activists alike have pointed the finger at lax security regulations and ever-growing tech monsters like Facebook for unethical and irresponsible corporate behaviour. And while there is no question that the Facebook leadership behaved capriciously with its users' data and allowed its continued growth to drive its incautious privacy policies—the question remains: When will we be proactive with our technology instead of reactive? The Cambridge Analytica scandal did not happen overnight; rather, it was a process that took shape over the span of years with our almost-complicit non-opposition. Certainly, the Zuckerbergs and the Nixes must take responsibility for their bottom-line driven management, but we need to develop preventative safeguards to preclude these disasters, especially considering that as our processing ability grows, its potential consequences will become increasingly detrimental.

Although we have seen how data can be harvested, stolen, sold and even abused, there are more visceral and potentially fatal consequences of unrestrained and under-regulated technological development. The quintessential example of this is the self-driving car that killed a woman crossing the street. In March, 2018, a woman in Tempe, Arizona, was struck and killed by a self-driving vehicle while she was crossing the street (Marshall, 2018). This crash represents yet another moment in which society realized the dangers of technology only after the damage had been done. The autonomous car was more than just a novelty, it was part of a burgeoning industry that stood to create fortunes (Olsen, 2018; Crichton, &Danny, 2018; Schleifer, 2019). As recently as a year ago, autonomous car startups were wrapping up monstrous fundraising series that seemed to promise a gilded future for the self-driving car industry. *Tech Crunch* reported in 2018 that infant startup Pony.ai raised USD 112 million for its autonomous car projects, while Zoox managed to amass USD 500 million in their series A and Aurora with secured USD 90 million in the same year. This funding boom, if anything, predicted a continued positive trend. However, the death of the Arizona woman, Elaine Herzberg, revealed the much bleaker side of experimental technology; and it seems that our awareness of the morbid potentialities of unchecked technological development came at the cost of human life. The police investigation carried out by the Tempe Police Department revealed that the Volvo XC90 SUV was in autonomous mode when it barreled into Herzberg (ibid., 32). Herzberg was evacuated from the scene and taken to a hospital where she later succumbed to her wounds. In response to the accident, Uber took all its self-driving cars off public roads and issued a statement acquiescing to full cooperation with the authorities. As US Senator Richard Blumenthal (D-Connecticut) remarked, 'This tragic incident makes clear that autonomous vehicle technology has a long way to go before it is truly safe for the passengers, pedestrians, and drivers who share America's roads' (ibid., 32).

Undoubtedly, one would think that such a catastrophic event in the nascent industry's development would encourage policy-makers to rethink the regulation of autonomous vehicles. However, the reality on the ground reveals a different picture. Before the tragic accident, Arizona Governor Doug Ducey signed a permissive executive order which afforded autonomous vehicle companies that the ability to 'test or operate fully driverless vehicles in the state' (Executive Order, 2018; ibid., 32). In California, where the regulation was slightly stricter, developers were mandated to keep a record of specific data. The numbers offered little insight into how safely the vehicles operate, but as of the end of October 2018, California permitted Waymo, the self-driving development firm owned by Alphabet, to test its cars on public roads, 'without a test driver present' (ibid., 32; Shoot, 2018; Team, 2018),

The death caused by Uber's self-driving car sparked a whole host of questions about the autonomous car industry. In particular, if the vehicles need to be tested on public roads, how ethical is permission to do so if pedestrians and other drivers on the road haven't consented to the experiment? Bryan Reimer, an expert who studies human behaviour and autonomous vehicles at MIT, summarized the nature of unrestrained tech perfectly: 'This is another major illustration that the technology we're talking about is evolving over time and not necessarily road ready for wide deployment' (ibid., 32).

The autonomous vehicle industry parades its self-driving cars as solutions to the deadly reality of motor vehicle fatalities. According to the Association For Safe International Road Travel, nearly 1.3 million people die in road accidents annually, while another 20–50 million people are injured or disabled (ibid., 32). The great purported boon of the self-driving car is that it could prevent thousands of car accidents caused by careless, drunk and tired drivers. However, the road-time of the autonomous car thus far is exponentially less than human-driven vehicles, and there has already been a fatality.

The objective of this book lies in filling the moral void for the public service to successfully negotiate the challenges created by AI technological innovations. This may encourage public administrators to be held responsible for the value-laden-ness of AI applications by making diverse social identities and public values visible as they can be critical players in considering multiple forms of disadvantage and discrimination.

The Book's outline

Chapter 1 shows how technological developments, such as algorithms, are designed to perform a task and take decisions autonomously with a particular moral delegation in mind. This moral delegation by designers affects the lives of individuals and shapes organizations and societies as a whole in terms of rights and dignity. This chapter focuses on the conceptualization of algorithms as value-laden, rather than neutral, as they are used in decision making to discuss their ethical implications. The omnipresent use of 'intelligent' algorithms in the public sector, with emphasis on AI technologies, requires greater ethical scrutiny and a relatively prudent approach regarding any implementation of AI-based technologies on a broad scale.

Chapter 2 depicts the multilayered conceptualization of AI and its various technological methods. The literature on AI is gaining more scholarly attention, due to the detrimental implications of AI usage in the public sphere. A more policy-oriented approach will be suggested to enhance the confrontation of the public sector with these technological obstacles and to maximize the full potential of public AI.

Chapter 3 presents existing approaches for responsible AI innovation governance. Responsible innovation, or RI, often refers to institutional structures for innovation that can lead to well-accepted technological advances.

This chapter aims to suggest ethical aspects of RI governance on society and the environment that can facilitate responsible decision-making by stakeholders.

Chapter 4 introduces the conceptual laying of prudential rationality ethics. This ethical notion brings space into the analysis of the application of machine learning, cognitive computing and algorithmic and data governance. Since space is relational, fluid and a constantly evolving concept, it should be used to contextualize self-regarding interests and other regarding interests, both of which have an equal part in AI management decisions, including: What are the implications of these constructions for how the technologies are presented? Who has agency in different constructions of public space and private space? How one can demarcate the boundaries between public and private spaces?

At the individual level, enlightened self-interest takes place when people act in ways to further the interests of others, which acts ultimately serve their own interests as well.

The junction of prudential ethics and intersectionality theory and AI allows us to conceptualize and harness, for the first time, patterns of inequality. To keep up with the technological pace while adhering to the public interest, the public sector should acquire the capacities, and consider overlapping social categories, for supporting and promoting social diversity and inclusion.

Chapter 5 carries out a critical analysis of current AI and big data affairs in the international arena, including Facebook's and Twitter's fake news scandals and their aligned government responses; Amazon's use of a secret AI recruiting tool that showed bias against women; mass surveillance systems enacted by governments, like the Chinese Social Credit and AI-based monitoring systems during the coronavirus pandemic; and more. Drawing on these technological affairs provides the opportunity to reexamine the professional role of public service and related claims of expertise, public interest, accountability and norms in the cognitive era.

References

Beckett, L. (2017). Trump digital director says Facebook helped win the White House. *The Guardian*. Retrieved from www.theguardian.com/technology/2017/oct/08/trump-digital-director-brad-parscale-facebook-advertising

Cadwalladr, C. (2017). The great British Brexit robbery: how our democracy was hijacked. *The Guardian*, 7.

Cadwalladr, C., & Graham-Harrison, E. (2018). Revealed: 50 million Facebook profiles harvested for Cambridge Analytica in major data breach. *The Guardian*, *17*, 22.

Crichton, D., & Danny C. (2018). One year old pony.ai raises $112 million series A to build autonomous car future. *TechCrunch*. Retrieved from https://techcrunch.com/2018/01/15/one-year-old-pony-ai-raises-112-million-series-a-to-build-autonomous-car-future/

Davies, H. (2015). Ted Cruz using firm that harvested data on millions of unwitting Facebook users. *The Guardian*, *11*, 2015.

Deagon, B. (2018). Facebook Stock Flashes Multiple Sell Signals Amid Cambridge Analytica Data Scandal | Stock News & Stock Market Analysis – IBD. *Investor's Business Daily*, 26 Mar. 2018.

Doward, J., Cadwalladr, C. & Gibbs, A. (2017). Watchdog to launch inquiry into misuse of data in politics. *The Guardian*. Retrieved from www.theguardian.com/technology/2017/mar/04/cambridge-analytics-data-brexit-trump

Executive Order. No. 2018-04 (2018), pp. 1–3. State of Arizona Executive Order; Governor Douglas A. Ducey.

Heilbroner, R. (1994). Technological determinism revisited. *Does technology drive history*, Smith M.R. & Marx L. 67–78. MIT Press.

Hewlett, R. G. & Anderson, O. (1962). The New World, 1939–1946. *University Park: Pennsylvania State University Press*.

Hewlett, R. G. & Anderson, Oscar E. (1962). *Groves' speech upon handing over control of the Manhattan Project to the Atomic Energy Commission*. The New World, 1939–1946 (PDF). Pennsylvania State University Press. ISBN 0-520-07186-7. OCLC 637004643. Retrieved 26 March 2013.

Lomas, N. (2018). How Facebook has reacted since the data misuse scandal broke. *Tech Crunch*. https://techcrunch.com/2018/04/10/how-facebook-has-reacted-since-the-data-misuse-scandal-broke/

Marshall, A. (2018). Uber's self-driving car just killed somebody. Now what? *Wired*, March 20, 2018.

Masco, J. (2013). *The nuclear borderlands: The Manhattan project in post-cold war New Mexico*. Princeton University Press.

Olsen, D. (2018). Self-Driving car startup launches amid funding boom. *PitchBook*.

Rosenberg, M., Confessore, N. & Cadwalladr, C. (2018). How Trump consultants exploited the Facebook data of millions. *New York Times*. Archived from the original on March 17, 2018.

Schleifer, T. (2019). Aurora, the hot self-driving startup, will be worth $2 billion after an investment by Sequoia. *Recode*.

Schwartz, M. (2017). Facebook failed to protect 30 million users from having their data harvested by Trump campaign affiliate. *The Intercept*. https://theintercept.com/2017/03/30/facebook-failed-to-protect-30-million-users-from-having-their-data-harvested-by-trump-campaign-affiliate/

Shoot, B. (2018). California permits Waymo to test driverless autonomous cars on its streets. *Fortune*, October 30, 2018.

Smith, M. R., & Marx, L. (Eds.). (1994). *Does technology drive history?: The dilemma of technological determinism*. MIT Press.

Team, W. (2018). A green light for Waymo's driverless testing in California. *Medium.com*, Waymo Blog, October 30, 2018. https://blog.waymo.com/2019/08/a-green-light-for-waymos-driverless.html

Watson, C. (2018). The key moments from Mark Zuckerberg's testimony to Congress. *The Guardian*, April 11, 2018. www.theguardian .com/technology/2018/apr/11/mark-zuckerbergs -testimony-to-congress-the-key-moments

1   Ethics of algorithms

Technological developments, such as algorithms, are designed to perform a task, to make decisions autonomously with a particular moral delegation in mind. This moral delegation by designers is affecting the lives of individuals and shaping organizations and societies as a whole in terms of rights and dignity. This chapter focuses on the conceptualization of algorithms used in decision-making as value-laden, rather than neutral, decision-making to discuss their ethical implications. The omnipresent use of "intelligent" algorithms in the public sector, with emphasis on AI technologies, requires greater ethical scrutiny and a relatively prudent approach regarding any implementation of AI-based technologies on a broad scale.

The role of algorithms in decision-making

In this section, we briefly introduce some key definitions of AI and Machine Learning (ML) in order to guide the reader through the chapter and then move on to an overview of machine ethics as a normative framework of the ethics of algorithms. As we expand algorithms' decision-making roles, ethical considerations are inevitable. With technological development, all of us—ethicists and designers included—are involved in evaluation processes requiring the selection and application of ethical standards. As such, algorithms must be treated as normative agents in the limited sense that we can assess how well they perform.

AI is the multidisciplinary attempt to build machines that can learn, make decisions, and act intelligently in the environment (Russell & Norvig, 2009). Decisions are the outputs of a learning process, as well as inputs to create machines.

Although there is no one accepted definition of AI, various scholars address the machine's ability to exhibit intelligent behaviour, react to the environment, and learn from it (Samoili et al., 2020). Nonetheless,

DOI: 10.4324/9781003106678-2

this chapter will focus on the AI's twofold functionality within the disinformation sphere. As such, AI is associated with complex algorithmic models that are capable to automatically generate, detect, and mitigate false content online and impact public opinion.

Machines are technological artifacts comprised of a combination of software and hardware components. ML is a discipline that combines statistical modeling and the science of algorithms to create computer systems able to automatically make effective predictions and support decision-making by learning inductively from input data (Mitchell, 1997; Vapnik, 2000). As Mitchell explains: "[e]ach ML problem can be precisely defined as the problem of improving some measure of performance P when executing some task T, through some type of training experience E" (Mitchell, 1997).

The potential power of AI in the cybernetic and physical spheres has led to a broad discourse regarding its ethical implications. On the one hand, AI-driven interfaces are able to analyse large amount of data and provide us with a tailored user experience and improved personalization processes, as can be seen within various fields such as autonomous driving, predictive policing, and language translation. On the other hand, there are indeterminate outcomes of AI usage from the legal, social, and ethical perspectives (Doshi-Velez et al., 2017; Amodei et al., 2016; Sculley et al., 2014; Bostrom, 2003; McCarthy, 1960).

The algorithm is viewed as a procedure implemented into computer-understandable platforms: "any well-defined computational procedure that takes some value, or set of values, as input and produces some value, or set of values, as output" (Cormen et al., 2001). An algorithm is able to transform language to generate the solution itself, by computing the parameters of the chosen ML model using available input data.

Positioning the algorithm as having an important role within the larger ethical decision-making requires a conceptualization of the basic properties of the general notion of the algorithm. A definition for algorithms has been given in disciplines such as computer science and mathematics. However, it should be noted that the concept of an algorithm cannot be precisely defined in full, at least for the time being. The reason for this is that the concept is expanding. Algorithms are treated as mathematical constructs: 'a finite, abstract, effective, compound control structure, imperatively given, accomplishing a given purpose under given provisions' (Hill, 2015, p. 36). This definition is given in accordance with Kleene's (1967) requirement that an algorithm should be defined in reference to a procedure that must be known in full before its deployment. As explained by Kleene (1967):

If (after the procedure has been described) we select any question of the class, the procedure will then tell us how to perform successive steps, after a finite number of which we will have the answer to the questions we selected. In performing the steps, we have only to follow the instructions mechanically, like robots; no insight or ingenuity or invention is required of us. After any step, if we don't have the answer yet, the instructions together with the existing situation will tell us what to do next. The instructions will enable us to recognize when the steps come to an end, and to read off from the resulting situation the answer to the question, 'yes' or 'no'. In particular, since no human performer can utilize more than a finite amount of information, the description of the procedure, by a list of rules or instructions, must be finite

(p. 223)

However, Hill (2015) also introduces a more popular definition of algorithms. Rather than a mathematical construct, in public discourse an algorithm is assessed to be more than its procedure, as a technological configuration in relation to a specific task or data set:

we see evidence that any procedure or decision process, however ill-defined, can be called an 'algorithm' in the press and in public discourse. We hear, in the news, of 'algorithms' that suggest potential mates for single people and algorithms that detect trends of financial benefit to marketers, with the implication that these algorithms may be right or wrong....

(Hill, 2015, p. 36)

Hill acknowledges the problem with the popular usage of algorithms is that it attempts to describe any procedure or decision-making process.

In this book, we focus on decision-making algorithms whose actions are difficult for humans to predict and are used across a variety of domains, from simplistic decision-making models (Levenson & Pettrey, 1994) to complex profiling algorithms (Hildebrandt, 2008). The extended scale of decision-making algorithms' usages across various domains thus raises ethical challenges and implications in the design and operation of algorithms with decision-making processes.

Key concepts of ethical algorithms

Existing literature discussing ethical aspects of algorithms identifies six key areas of concern resulting from the operation of algorithms.

Mittelstadt et al. (2016) focus on three main aspects of algorithms' operation including: (a) the way algorithms turn data into evidence for a given outcome (producing conclusion), and that this outcome is then used to (b) activate and motivate an action that may be value-laden (Brey & Soraker, 2009; Wiener, 1988). This operation of algorithms producing evidence also raises issues of responsibility and accountability (c) for effects of actions driven by algorithms. The organizing structure offered by Mittelstadt (2016) on the operation of algorithms allows building a normative framework of ethical issues arising from the use of algorithms, based on how algorithms process data to produce evidence and motivate actions. Thus, building a normative framework should be based on six key areas of ethical aspects of algorithms including inconclusive evidence, inscrutable evidence, misguided evidence, unfair outcomes, transformative effects, and traceability.

Inconclusive evidence

Algorithms are capable of drawing conclusions from the data they process using inferential statistics and/or machine learning techniques for creating probable knowledge in terms of statistical methods to quantify levels of probability (Mittelstadt et al., 2016; James et al., 2013; Hacking, 2006). Statistical learning methods and computational learning theories are also used to identify significant correlations despite its inability to bring the ultimate sufficiency of existing causal connections (Illari & Russo, 2014), and thus may be insufficient to motivate action on the basis of knowledge from such a connection. These limitations are echoed in actionable insights. Actionable insights are generated by analyzing information and drawing conclusions that give a predictive insight into certain actions that should be taken by decision-makers. Leveraging data analytics to provide actionable insights with algorithms when more reliable devices are either not available or too costly to implement, may be unreliable. Acknowledging this drawback is important and should be accompanied by a risk assessment for harms to one's epistemic responsibilities (Miller & Record, 2013). In other words, the justification for a certain action should be based on an acceptable conclusion drawn from given data analysis (Grindrod, 2014).

Inscrutable evidence

Scrutiny of data processing and assessing its credibility is important when drawing conclusions, make inferences, or suggesting implications based on specific data and evidence (Mittelstadt et al., 2016). When the casual connection is not reliable, this expectation can be satisfied by better access as well as by additional explanations. Thus, algorithms processing data may cause serious limitations for drawing conclusions by a machine-learning algorithm. (Miller & Record, 2013)

Misguided evidence

Algorithms process data and therefore may suffer from epistemic problems concerning the nature of algorithmic evidence. Shannon's mathematical theory of communication (Shannon & Weaver, 1998) demonstrates this aspect by the informal 'garbage in, garbage out' principle for alleviating the misguided and inconclusive evidence as neutral and not observer-dependent.

Unfair outcomes

Actions driven by algorithms are often prone to biases. As such, AI researchers have to produce algorithms that avoid, or at least minimise, unfair or discriminatory actions (Mittelstadt et al., 2016). However, avoiding unfair outcomes maybe to the cost of compromising accuracy. Aiming for fairness outside of the algorithm's design would initial observer dependent's fairness of the action and its effects, thus leaving algorithm designers free to focus on maximising accuracy, with fairness left to state regulators, with expert and democratic input (Newman & Harmon, 2016).

Transformative effects

The ethical challenges posed by the reliance on algorithmic data-processing and (semi-) autonomous decision-making emerge from the new roles that algorithmic activities, like profiling, take in social functions. Algorithms are able to affect how people conceptualize the world and thus shape the ways people can conceptualize the fairness and trustworthiness of algorithmic decisions on preferred actions (Lee & Baykal, 2017). While it is important to ensure that algorithms make fair and trustworthy decisions, more critical is to explore the acknowledged contexts in which algorithmic decisions are embedded.

Traceability

Since algorithms are software-artifacts used in data processing, and as such are dependent on the design and availability of new technologies and those associated with the manipulation of large volumes of personal and other data, the responsibility for algorithms in decision-making is not clear. While designers are assumed to hold responsibility for their algorithms later in use, the companies and other stakeholders in the algorithms creation also hold that responsibility (Kroll et al., 2017). If one considers algorithms as value-laden, rather than neutral, it is important to identify who should be held accountable for the ethical implications of the algorithm in use. However, it may not be straightforward to trace who should be held responsible for the harm caused. (Neyland, 2016; Martin, 2019)

This brief review of both epistemic and ethical challenges raised by algorithms and evidence produced by them allows demarcating widely interdisciplinary academic discourse on the ethical aspects of algorithms (Mittelstadt et al., 2016). Drawing on the key areas of the ethical aspects of algorithms, a fundamental rethinking of ethics, citizenship, and democracy and what should be the normative grounding for responsibility for algorithms when in use is needed. We now turn to the impetus for demonstrating how ethics for algorithms enhance individual and collective well-being.

Ethical implications of algorithmic design

As seen in the previous section, the ethics of algorithms have evolved around six key areas of ethical aspects including; inconclusive evidence, inscrutable evidence, misguided evidence, unfair outcomes, transformative effects, and traceability (Mittelstadt et al., 2016). The following sections demonstrate how the ethical implications of algorithmic design and use are discussed in the existing literature.

Algorithmic inconclusive evidence guides unjustified actions

Critiques of algorithms' role in decision-making and data mining often argue that causal links and correlations identified within a dataset may be problematic although they are seen as sufficiently credible to direct action without first establishing causality (Hildebrandt, 2011; Hildebrandt & Koops, 2010; Mayer-Schönberger & Cukier, 2013; Zarsky, 2016). Ananny (2016) explains that efficient and scalable systems require stable categories of individuals who use the same phrases and act in predictable ways. Thus, machine learning algorithms require a huge amount of data before they are able to reliably affect decisions or recommended actions. As Ananny has critiqued, the predictive analytics correlations established in large (2016: 103), 'algorithmic categories…

signal certainty, discourage alternative explorations, and create coherence among disparate objects,' all of which affect the way individuals are being described (possibly inaccurately) or influence communities via simplified models or classes (Barocas, 2014).

Following Ananny (2016), algorithmic inconclusive evidence has an ethical power by building a disciplined community of humans and machines that resembles and recreates probabilities, to suggest what possible outcomes the model anticipates likely and reasonable (Mackenzie 2015).

Transparency and interpretability of algorithms is infeasible

The arguments for and against transparency as a design requirement of machine learning algorithms has become a major ethical concern in the reviewed literature (Tutt, 2016; Crawford, 2016; Neyland, 2016; Raymond, 2014).

As for the term transparency, it is generally defined as '[t]he availability of information, the conditions of accessibility and how the information...may pragmatically or epistemically support the user's decision-making process' (Turilli & Floridi, 2009: 106). As such, transparency refers to data available about an algorithm that specifies part of its decision-making process or data about the decisions it makes, which can be taken by a human being. Drawing on this conceptualization, an accessor of the data could be a data scientist, policy maker, or even a citizen. It should be added that the component of interpretability is a key requirement here, ensuring that published data does actually enhance our understanding of algorithmic processes (Ananny & Crawford, 2016). Ethical concerns about the fairness of algorithms have increased the call for machine learning methods that are transparent. This growing normative purchase of transparency of algorithms involves two types of transparency associated with fairness, namely, a procedural aspect of transparency (how much we comprehend the internal state of an algorithm) and outcome transparency (how much we understand about the decisions, and forms in decisions, made by an algorithm). A critical assessment of the algorithm's transparency is important especially when it comes to harmful/obstructive uses of transparency (i.e., it is argued that many machine learning methods are useful precisely because they work in a way which is alien to conscious human reasoning, for the purpose of ensuring competitive advantage or securing privacy)(Glenn & Monteith, 2014; Kitchin, 2016; Stark & Fins, 2013; Leese, 2014).

In addition to having adverse consequences or not following the principles of fairness, making algorithms transparent dictates appropriate levels of accessibility and comprehensiveness (Turilli & Floridi, 2009). Producing an interpretable explanation without knowing anything about how the algorithm operates lends to the portrayal of machine learning algorithms as a 'black box'. According to Burrell (2016) algorithms 'are opaque in the sense that if one is a recipient of the output of the algorithm (the classification decision), rarely does one have any concrete sense of how or why a particular classification has been arrived at from inputs' (Burrell, 2016, p. 1).

It may be helpful to frame this issue in the context of the alternative, which is human-led decision making. Since decisions made by people can occasionally be opaque and susceptible to bias then human intervention in algorithmic decision-making '[i]s impossible when the machine has an informational advantage over the operator…[or] when the machine cannot be controlled by a human in real-time due to its processing speed and the multitude of operational variables' (Matthias, 2004, pp. 182–183).

Ultimately, algorithmic processing can be distinguished from traditional decision-making in that human decision-makers can, in principle, articulate their rationale when queried, limited only by their desire and capacity to provide an explanation, and the questioner's capacity to understand it (Mittelstadt et al., 2016). Under these conditions, it is difficult to analyse the inner workings of particular techniques in a way that will not slow or prevent their uptake. A realistic approach would be to invest in data processors and controllers to ensure a trusting relationship with data subjects. (Cohen et al., 2014; Shackelford & Raymond, 2014) Building trust in machine learning techniques must reframe algorithmic decision-making to include ethical implications rendering the choice of factors, sourcing the data and assessment of the output with data subjects.

Biased decisions by algorithms

Algorithmic decision-making delivers on the promise that algorithms work in a way that is alien to conscious human reasoning (Bozdag, 2013; Naik & Bhide, 2014). However, scholars often argue that algorithms are constructed by individuals in development, design, implementation, and use, thus making algorithms value-laden with preferences for certain outputs (Bozdag, 2013; Friedman & Nissenbaum, 1996; Kraemer et al., 2011; Macnish, 2012; Newell & Marabelli, 2015). For that bias manifests in algorithms and the outcomes they produce so that 'the values of the author [of an algorithm], wittingly or not, are frozen into the code, effectively institutionalising those values' (Macnish, 2012: 158).

Algorithmic development processes are not necessarily devoid of societal biases, prejudice, stereotypes, and even incorrect assessment about individuals and their experiences. Friedman and Nissenbaum's (1996) work underscored the ethical implications raised by computer systems which contain biases embedded in '[s]ocial institutions, practices and attitudes' as well as technical bias that can occur due to technical limitations and emergent bias which appears sometime after software implementation is completed and applied (Friedman & Nissenbaum, 1996).

Preexisting bias is rooted in social institutions, practices, and attitudes. In computer systems social biases often exist prior to the creation of the system which is embedded in subcultures, and in formal or informal practices, across organizations and institutions. They can also be manifested by the personal biases of system designers or clients. Personal bias may be created through explicit and mindful efforts of individuals or institutions, or implicitly and unconsciously. For example, the design of machine learning algorithms is affected by choices made by designers, i.e. which factors to include in the algorithm and how to weigh them. To serve majority interests, information intermediaries often embed a popularity metric in their ranking algorithm (Bozdag, 2013).

Friedman and Nissenbaum (1996) argue that technical bias arises from technological constraints, errors or design decisions, which favour particular groups without an underlying rationale. Examples include when an alphabetical listing of airline companies leads to increase business for those listed earlier in the alphabet, or an error in the design of a random number generator that causes particular numbers to be ranked higher. Machine learning biases can exhibit bias that manifest in the datasets processed by algorithms. Flaws in the data are unintentionally chosen by the algorithm and veiled in outputs and models produced (Romei & Ruggieri, 2014).

Emergent bias can be hard to identify since it arises only in a context of use, as a result of changing societal knowledge, population, or cultural values (Friedman & Nissenbaum, 1996). Emergent bias exhibits decisional rules developed by the algorithm, rather than any 'hand-written' decision-making structure (Mittelstadt et al., 2016). The algorithm's output is likely to be prone to emergent bias because interfaces by design seek to reflect the capacities, character, and habits of prospective users. This reflects the interpreter's 'unconscious motivations, particular emotions, deliberate choices, socio-economic determinations, geographic or demographic influences' (Hildebrandt, 2011, p. 376).

Profiling algorithms leading to discrimination

Algorithmic profiling is defined as a method of using the inferential analysis to identify correlations or patterns within datasets, that become an indicator to make predictions about behaviour of a group's membership (Hildebrandt, 2008; Schreurs et al., 2008). According to Hildebrandt and Koops (2010), the fact that '[t]he construction or inference of patterns by means of data mining and...the application of the ensuing profiles to people whose data match with them' (p. 431) can lead to discrimination. Algorithmic profiling may result in social sorting and other discriminatory outcomes since the categories are established from 'probabilistic assumptions' (Leese, 2014, p. 502) that are de-individualized (Schermer, 2013; Lyon, 2003, 2014; Parsons, 2015). For example, in research studies in Australia (Mann & Daly, 2019) and North America (Browne, 2015; Eubanks, 2018; Noble, 2018; Sandvig et al., 2016) scholars showed that algorithmic profiling targets marginalised groups, including racial groups, low socio-economic status groups, and women. Browne (2015) argues that the origins can be traced to datasets that are constructed that disproportionately contain data about certain individuals, leading to over monitoring and over-policing of those groups (Ferguson, 2017). As such, algorithmic profiling disseminates hierarchies predicated on the entangling of identity characteristics.

Following Schermer's (2011), discriminatory practice toward specific individuals and communities is not ethically problematic in itself; rather, it is the effects of the treatment that determine its ethical acceptability. However, Schermer uses bias and discrimination interchangeably. For Schermer, bias is a component of the decision-making itself, whereas discrimination describes the effects of a decision, in terms of adverse consequences resulting from algorithmic decision making. Barocas and Selbst (2016) suggest that incongruent impact detection provides a model for the detection of bias and discrimination in algorithmic decision-making which is sensitive to differential privacy. Bias may be detected in the beginning stages of the algorithm's development and execution by avoiding sensitive characteristics that contribute to discrimination (Barocas and Selbst, 2016), such as gender, race or ethnicity (Kamiran & Calders, 2010; Schermer, 2011), depending on the particular context that carries discriminatory potential.

Removing or reducing unwanted discrimination in the model-building process is still insufficient, particularly since algorithmic profiling established from neutral characteristics such as postal code may inadvertently intersect with other profiles related to ethnicity, gender, sexual preference, etc. (Macnish, 2012; Schermer, 2011). Given the limits on certain training data, Romei and Ruggieri (2014) offer four overlapping strategies for discrimination prevention in analytics: (1) monitoring misrepresentation of training data; (2) integration of anti-discrimination criteria into the classifier algorithm; (3) post-processing of classification models; (4) adjustment of predictions and decisions to ensure a fair proportion of effects for both protected and unprotected groups (Mittelstadt et al., 2016, p. 8). These strategies can contribute to algorithmic fairness as it considers not only discrimination but fairness, neutrality, and independence as well (Kamishima et al., 2012). On this point, a discussion of algorithmic fairness seems appropriate. Among data scientists, metrics of fairness are suggested based on statistical parity, differential privacy and other relations between data subjects in classification tasks (Dwork et al., 2012; Romei & Ruggieri, 2014). The practice of personalization is offered to ensure that algorithmic filtering maintains to 'replacing the traditional repositories that individuals and organizations turn to for the information needed to solve problems and make decisions.' (Mowshowitz & Kawaguchi 2002). Personalisation through non-distributive profiling allows increasing user collaboration and user generated content however, the capacity of individuals to investigate the personal relevance of factors used in decision-making is inhibited by opacity and automation (Zarsky, 2016).

Algorithm's threat to autonomy

Personalizing algorithms affect the behaviour of data subjects and human decision makers by filtering information that is closely related to autonomy. The ethical dilemma of autonomy defined by Brey (2000) as '[s]elf-governance, that is, the ability to construct one's own goals and values, and to have the freedom to make choices and plans and act in ways that are believed by one to help achieve these goals and promote these values' (p. 14).

Drawing on Brey's conceptualization, to be self-governing and make choices one needs to be properly informed. The unprecedented accessibility of data offered by the Internet is seen as pivotal to individuals' autonomy. However, the large amount of information available requires filtering which in turn reduces user autonomy. The role of personalization algorithms provides a gatekeeping function in the cost of value-laden decisions affecting the autonomy of data subjects in order to align it to users' personal preferences (Annay, 2016; Bozdag & Timmermans, 2011).

The value of autonomy implies that the information available to the user should be based upon an in-depth assessment of preferences, behaviours, and vulnerabilities of individuals rather than third party interests (Newell & Marabelli, 2015; Zarsky, 2016; Applin & Fischer, 2015; Stark & Fins, 2013; Bozdag, 2013; Goldman, 2006). While mitigating challenges to autonomy, computer and data scientists demonstrate that personalisation improves decision-making by providing the subject with only relevant information when met with a huge amount of data. According to Johnson, the data subject is forced to make the 'institutionally preferred action rather than their own preference' (Johnson, 2013) so that personalisation algorithms need to be trained to 'act ethically' in reaching a balance between coercing and supporting users' autonomy in decision making (Lewis & Westlund, 2015). This strategy is associated with Sunstein (2007) termed as 'echo chambers'. For Sunstein, citizens would use technological tools to devoid contradictory information that impedes decisional autonomy, leading to what he defines as 'echo chambers' or 'information cocoons'. It should further be noted that the value of autonomy is potentially in conflict with information diversity (Bozdag, 2013; Raymond, 2014).

Challenges for informational privacy

Informational privacy is an ethical and legal issue associated with profiling and data mining, securing that data subjects have the right and the ability to protect their personal data from third parties (Schermer, 2011, p. 48). The right to an identity derived from informational privacy interests suggests that secretive profiling is of great concern (Van Wel & Royakkers, 2004; Mittelstadt et al., 2016).

Decision-making driven by algorithms leads to the incapacity of data subjects to define privacy norms to govern all types of data generically because their value is only established through processing (Hildebrandt, 2011; Van Wel & Royakkers, 2004). Developing effective privacy protection technologies is a critical challenge to opacity and managing identity by analytics techniques. Schermer (2011) identifies the challenge raised by informational privacy as a normative foundation for algorithmic decision-making as

> Informational privacy does not only decrease the usefulness of data-mining, it also increases the probabilities of false positives and false negatives. Moreover, data exclusion and data minimisation do not necessarily provide sufficient protection from the risks of profiling,

because it is often still possible to uniquely identify particular persons from the data, even after key attributes such as name, address and social security number have been removed. This indirect identification is troublesome from a legal perspective, since personal data protection law is so dependent on the notion of personal data. Also, it is difficult to determine which pieces of data can lead to the identification of a person once combined.

(Schermer, 2011, p. 49)

This weakness in informational privacy lies in the fact that it is based primarily on ex ante protection, but has little in the way of ex post protection mechanisms (Schermer, 2011).

A solution would be to create profiling through web-data mining thus leading to de-individualisation, identifying individuals based on the basis of group characteristics instead of on their own individual characteristics and merits (Van Wel & Royakkers, 2004, 133). In terms of data accuracy, individuals will be judged and treated as group members rather than individuals and therefore will be at risk of stigmatization just by being labeled as a member of a certain group (Van Wel & Royakkers, 2004). Thus, the individual's informational identity (Floridi, 2011) may be threatened by algorithms that link the subject to others within a dataset (Vries, 2010). Other mechanisms for data subjects can be an 'optout' of profiling for a particular purpose or context that would help protect data subjects' privacy interests (Hildebrandt, 2011; Rubel & Jones, 2014).

As noted by Hildebrandt and Koops (2010) a strategy of 'smart transparency' can better address the challenges posed by informational privacy by designing the socio-technical infrastructures responsible for profiling in a way that allows deliberation and ensures collaboration of individuals in the process of profiling.

Traceability as a path to moral responsibility

One of the ethical dilemmas in machine ethics deals with the question of to what extent is it possible to hold the manufacturer/programmer/user of an autonomous, learning automaton accountable for the actions of the machine? Ascribing responsibility for algorithms is not clear. In the context of machine ethics, if a machine does not operate according to the manufacturer's operating scheme, we hold the responsibility to the manufacturer of the machine instead of the user. Thus, the necessary condition of responsibility lies in control. Blameworthiness can only be justifiably attributed when the actor has some degree of control (Matthias, 2004) and intentionality in carrying out the action:

> The operating manual of a device transfers control of that device from the manufacturer to the operator, by specifying the precise set of actions and reactions (in a system theory vocabulary: of transformations) the device is expected to undergo during normal operation, thus enabling the operator to handle it in a predictable manner, according to her own decisions on how to act.
>
> (Matthias, 2004, pp. 175–176)

This traditional conception of the delegation of responsibility in software design undertakes the programmer as the actor who has control over the technology's likely effects and potential for malfunctioning (Floridi et al., 2014) by preferencing desirable outcomes according to the functional specification (Matthias, 2004).

Delegation of responsibility appears to be inevitable once **algorithms** are viewed as holding powerful control over decision-making. Martin (2019) suggests the following: '[t]he more the algorithm is constructed as inscrutable and autonomous, the more accountability attributed to the algorithm and the firm that designed the algorithm' (p. 844). Such a determinist approach ascribes responsibility to algorithms on the basis of their capacity to enforce morality by preferring specific outcomes. As Desai and Kroll (2017) argue

> Some may believe algorithms should be constructed to provide moral guidance or enforce a given morality. Others claim that moral choices are vested with a system's users and that the system itself should be neutral, allowing all types of use and with moral valences originating with the user. In either case,…the author's deference to algorithms is a type of 'worship' that reverses the skepticism of the Enlightenment. Asking algorithms 'to enforce morality' is not only a type of idolatry, it also presumes we know whose morality they enforce and can define what moral outcomes are sought.
>
> (p. 118)

Morek (2006) is skeptical of the idea that algorithms can replace skilled professionals who have implicit knowledge and subtle skills that can hardly be made computable.

Desai and Kroll (2017) identify the drawback in the traditional criteria of control for responsibility delegation. Such criteria cannot expose less visible or completely overlooked outcomes and other actors' roles in decision making. This failure results from the mere fact that 'nobody has enough control over the machine's actions to be able to assume the responsibility for them' (Matthias, 2004, p. 177). As demonstrated by Martin (2019),

> the COMPAS algorithm was designed to preclude individuals from understanding how it works or from taking any responsibility for how it is implemented. Importantly, this is a design choice because other risk assessment algorithms are designed to be more open, thereby delegating more responsibility for the decision to individuals.

Therefore, the moral capacity of algorithms within decision-making remains controversial and leads to missed opportunities for investigating algorithms' responsibility once it is analysed in terms of control over decision making (Allen et al, 2006; Anderson, 2008; Floridi & Sanders, 2004). Allen et al. (2006, p. 14) seeks to offer a form of machine ethics that examines a necessary perquisite for fully exploring algorithmic responsibility: 'the modular design of systems can mean that no single person or group can fully grasp the manner in which the system will interact or respond to a complex flow of new inputs.'

It is suggested that for learning algorithms, responsibility can benefit from the concept of moral agency. Floridi and Sanders (2004) and Sullins (2006) argue that assigning moral agency to a machine brings to the fore issues of autonomy, interactive behaviour, and a role with causal accountability. Denying moral agency to machines carries the need to hold designers responsible for the unethical behaviour of their products (Anderson & Anderson, 2015; Kraemer et al., 2011; Turilli, 2007).

Calls for assigning responsibility to designers entrenched in Friedman and Nissenbaum (1996) suggestion that **developers** have a responsibility to design for diverse contexts ruled by different moral frameworks. As such, Turilli (2007) offers collaborative development of ethical requirements for computational systems to lay an operational ethical protocol. Since algorithmic data crosses context and decisions, consistency between the protocol (consisting of a decision-making structure) and the designer's or organisation's explicit ethical principles are pivotal (Turilli & Floridi, 2009).

At the other extreme, accountability may fall exclusively on the **users or on the society** as a whole, once algorithms are viewed as value-laden actors within decisions (Bozdag, 2013). **Developers** are able to design the algorithm to allow users to take responsibility for algorithmic decisions. The rationale behind this approach is that developers may inscribe the algorithm with the value-laden biases as well as roles and responsibilities of the algorithmic decision making. Users are assumed to be kept 'in the loop' of automated decision-making by algorithm's designers, thus remaining poorly equipped to identify ethical implications for algorithmic decisions and take corrective actions (Elish, 2019).

For that, a more holistic oversight of algorithmic decision-making pathways and dependencies is suggested for demarcating a line between algorithmic and human decision makers roles so that norms are required to guide when and how the human intervention is required, particularly in cases where real-time intervention is impossible before harms occur (Davis et al., 2013; Raymond, 2014). This provides a compelling basis for bridging the accountability gap between the designer's control and the algorithm's behaviour (Cardona, 2008) wherein blame can potentially be ascribed to several moral agents simultaneously.

Finally, algorithms impact the way individuals have access to social goods and rights, and the critical role they play in decision-making across social domains. Acknowledging the value-laden role of algorithms in decisions, raises the need to provide a normative grounding for the ethics of algorithms. In this chapter, we reviewed contemporary literature on the ethical challenges related to the transformative effects and traceability of algorithms. The transformative effect of algorithms leads to changes in identity construction and to issues of privacy and data protection mechanisms that operate at both individual and group levels. Traceability effects pose issues of accountability and the moral responsibility of algorithmic decisions. Issues of accountability and moral responsibility should be framed within a complex network of human and algorithmic actors.

References

Allen, C., Wallach, W., & Smit, I. (2006). Why machine ethics? *IEEE Intelligent Systems, 21*(4), 12–17.

Amodei, D., Olah, C., Steinhardt, J., Christiano, P., Schulman, J., & Mané, D. (2016). *Concrete problems in AI safety*. arXivLabs at Cornell University. *arXiv preprint arXiv:1606.06565*. Available at https://arxiv.org/abs/1606.06565

Ananny, M. (2016). Toward an ethics of algorithms: Convening, observation, probability, and timeliness. *Science, Technology, & Human Values, 41*(1), 93–117.

Ananny, M., & Crawford, K. (2016). Seeing without knowing: Limitations of the transparency ideal and its application to algorithmic accountability. *New Media & Society, 18*(3), 373–390.

Anderson, S. L. (2008) Asimov's 'Three Laws of Robotics' and machine metaethics. *AI and Society, 22*(4), 477–493.

Anderson, M., & Anderson, S. L. (2015). Towards ensuring ethical behaviour from autonomous systems: A case-supported principle-based paradigm. In *Artificial intelligence and ethics: Papers from the 2015 AAAI Workshop* (pp. 1–10).

Applin, S. A., & Fischer, M. D. (2015). New technologies and mixed-use convergence: How humans and algorithms are adapting to each other. In *2015 IEEE international symposium on technology and society (ISTAS)* (pp. 1–6). IEEE.

Barocas S. (2014). Data mining and the discourse on discrimination. Available at: https ://dataethics.github.io/proceedings/DataMiningandtheDiscourseOnDiscrimination.pdf

Barocas, S., & Selbst, A. D. (2016). Big data's disparate impact. *California Law Review, 104(3)*, 671–732.

Bostrom, N. (2003). 'Ethical issues in advanced artificial intelligence'. *Science fiction and philosophy: from time travel to superintelligence*, 277–284. Available at www.fhi.ox.ac.uk/publications/bostrom-n-2003-ethical-issues-in-advanced-artificial-intelligence-science-fiction-and-philosophy-from-time-travel-to-superintelligence-277-284/

Bozdag, E. (2013) Bias in algorithmic filtering and personalization. *Ethics and Information Technology, 15*(3), 209–227.

Bozdag, E., & Timmermans, J. F.C. (2011). Values in the filter bubble Ethics of Personalization Algorithms in Cloud Computing. Paper presented at the 1st International Workshop on Values in Design-Building Bridges between RE, HCI and Ethics, Lisbon, Portugal, September 2011, 7–15. Available at http://mmi.tudelft.nl/ValuesInDesign11/proceedings.pdf. Proceedings 1st International.

Brey, P., & Søraker, J. H. (2009). Philosophy of computing and information technology. In A. Meijers (Ed.), *Philosophy of technology and engineering sciences* (pp. 1341–1407). North-Holland.

Brey, P. (2000). Disclosive computer ethics. *Computers and Society, 30*(4), 10–16.

Browne, S. (2015). *Dark matters: On the surveillance of blackness*. Duke University Press.

Burrell, J. (2016). How the machine 'thinks:' Understanding opacity in machine learning algorithms. *Big Data & Society, 3*(1), 1–12.

Cardona B. (2008) 'Healthy ageing' policies and anti-ageingideologies and practices : On the exercise of responsibility. *Medicine, Health Care and Philosophy, 11*(4), 475–483.

Cohen, I. G., Amarasingham, R., Shah, Xie, B., & Lo, B. (2014). The legal and ethical concerns that arise from using complex predictive analytics in health care. *Health Affairs, 33*(7), 1139–1147.

Cormen, T. H., Leiserson, C. E., & Rivest, R. L. (2001). *Introduction to algorithms* (2nd ed.). MIT Press.

Desai, D. R., & Kroll, J. A. (2017). Trust but verify: A guide to algorithms and the law. *Harvard Journal of Law & Technology, 31*, 1–64.

Elish, M. C. (2019). Moral crumple zones: Cautionary tales in human-robot interaction. *Engaging Science, Technology, and Society, 5*, 40–60.

Eubanks, V. (2018). *Automating inequality: How high-tech tools profile, police and punish the poor*. St Martin's Press.

Ferguson, A. G. (2017). *The rise of big data policing: Surveillance, race and the future of law enforcement*. NYU Press.

Friedman, B. & Nissenbaum, H. (1996). Bias in computer systems. *ACM Transactions on Information Systems (TOIS), 14*(3), 330–347.

Glenn, T., & Monteith, S. (2014). New measures of mental state and behaviour based on data collected from sensors, smartphones, and the internet. *Current Psychiatry Reports, 16*(12), 1–10.

Goldman, D. (2006). *Social intelligence: The new science of human relationships*. Bantam Books.

Crawford, K. (2016). Can an algorithm be agonistic? Ten scenes from life in calculated publics. *Science, Technology & Human Values, 41*(1), 77–92.

Davis. M., Kumiega, A., & Van Vliet, B. (2013). Ethics, finance, and automation: A preliminary survey of problems in high frequency trading. *Science and Engineering Ethics, 19*(3), 851–874.

de Vries, K. (2010). Identity, profiling algorithms and a world of ambient intelligence. *Ethics and Information Technology, 12*(1), 71–85.

Doshi-Velez, F., Kortz, M., Budish, R., Bavitz, C., Gershman, S., O'Brien, D., Shieber, S., Waldo, J., Weinberger, D., & Wood, A. (2017). 'Accountability of AI under the law: The role of explanation'. arXivLabs at Cornell University. *arXiv preprint arXiv:1711.01134*.

Dwork, C., Hardt, M., Pitassi, T., Reingold, O., & Zemel, R. (2012). Fairness through awareness. In *Proceedings of the 3rd innovations in theoretical computer science conference* (pp. 214–226). ACM.

Floridi, L. (2011). The informational nature of personal identity. *Minds and Machines, 21*(4), 549–566.

Floridi, L., Fresco, N., & Primiero, G. (2014). On malfunctioning software. *Synthese, 192*(4), 1199–1220.

Floridi, L., & Sanders, J. W. (2004). On the Morality of Artificial Agents. Minds and Machines : Journal for Artificial Intelligence, *Philosophy and Cognitive Science,* 14(3), 349–379.

Grindrod, P. (2014). *Mathematical underpinnings of analytics: Theory and applications*. OUP.

Hacking, I. (2006). *The emergence of probability: A philosophical study of early ideas about probability, induction and statistical inference*. Cambridge University Press.

Hildebrandt, M. (2011). Who needs stories if you can get the data? ISPs in the era of big number crunching. *Philosophy & Technology*, 24(4), 371–390.

Hildebrandt, M. & Koops, B. J. (2010). The challenges of ambient law and legal protection in the profiling era. *The Modern Law Review, 73*(3), 428–460.

Hildebrandt, M. (2008). Defining profiling: A new type of knowledge? In: Hildebrandt, M. and Gutwirth, S. (Eds.), *Profiling the European citizen: cross-disciplinary perspectives* (pp. 17–45). Springer.

Hill, R. K. (2015). What an algorithm is. *Philosophy & Technology, 29*(1), 35–59.

Kamiran, T., & Calders, T. (2010). Classification with no discrimination by preferential sampling. *In Proceedings of the 19th machine learning conference (Belgium and the Netherlands, Leuven, Belgium).* Morgan Kaufmann.

Kamishima T., Akaho S., Asoh H., & Sakuma J. (2012) Fairness-aware classifier with prejudice remover regularizer. In *Joint European Conference on Machine Learning and Knowledge Discovery in Databases,* (pp. 35–50). Springer.

Kitchin, R. (2016). Thinking critically about and researching algorithms. *Information, Communication & Society, 20*(1), 14–29.

Kleene, S. C. (1967). *Mathematical logic.* Wiley.

Kraemer, F., van Overveld, K., & Peterson, M. (2011). Is there an ethics of algorithms? *Ethics and Information Technology, 13*(3) 251–260.

Kroll, J. A., Huey, J., Barocas, S., Felten, E. W., Reidenberg, J. R., Robinson, D. G., & Yu, H. (2017). Accountable algorithms. *University of Pennsylvania Law Review, 165.* https://papers.ssrn.com/sol3/papers.cfm?abstract_id=2765268.

Illari, P. M., & Russo, F. (2014). *Causality: Philosophical theory meets scientific practice.* Oxford University Press.

James, G., Witten, D., Hastie, T.& Tibshirani R. (2013). *An introduction to statistical learning.* Vol. 6, Springer.

Johnson, J. A., (2013). Ethics of data mining and predictive analytics in higher education (SSRN Scholarly Paper No. ID 2156058). Social Science Research Network.

Lee, M. K., & Baykal S. (2017). Algorithmic mediation in group decisions: Fairness perceptions of algorithmically mediated vs. discussion-based social division. *In Proceedings of the ACM Conference on Computer-Supported Cooperative Work & Social Computing (CSCW 2017),* (pp. 1035–1048). Association for Computing Machinery.

Leese, M. (2014). The new profiling: Algorithms, black boxes, and the failure of anti- discriminatory safeguards in the European Union. *Security Dialogue, 45*(5), 494–511.

Levenson, J.L., & Pettrey, L. (1994). Controversial decisions regarding treatment and DNR: An algorithmic Guide for the Uncertain in Decision making Ethics (GUIDE). *American Journal of Critical Care: An Official Publication, American Association of Critical-Care Nurses, 3*(2), 87–91.

Lewis, S. C., & Westlund, O. (2015). Big data and journalism. *Digital Journalism, 3*(3), 447–466.

Lyon, D. (2003). Surveillance as social sorting: Computer codes and mobile bodies. In D. Lyon (Ed.), *Surveillance as social sorting: Privacy, risk, and digital discrimination.* Routledge.

Lyon, D. (2014). Surveillance, snowden, and big data: Capacities, consequences, critique. *Big Data & Society, 1*(2), 1–13.

Mackenzie, A. (2015). The Production of Prediction: What Does Machine Learning Want? *European Journal of Cultural Studies, 18* (4–5), 429–45.

Macnish, K. (2012). Unblinking eyes: The ethics of automating surveillance. *Ethics and Information Technology, 14*(2), 151–167.

Mann, M., & Daly, A. (2019). (Big) data and the North-in-South: Australia's informational imperialism and digital colonialism. *Television & New Media, 20*(4), 379–395.

Martin, K. (2019). Ethical Implications and Accountability of Algorithms. *Journal of Business Ethics, 160*, 835–850.

Matthias, A. (2004). The responsibility gap: Ascribing responsibility for the actions of learning automata. *Ethics and Information Technology, 6*(3), 175–183.

McCarthy, J. (1960). *Programs with common sense* (pp. 300–307). RLE and MIT computation center.

Mayer-Schönberger, V., & Cukier, K. (2013). *Big data: A revolution that will transform how we live, work, and think.* Houghton Mifflin Harcourt.

Miller, B., & Record, I. (2013). Justified belief in a digital age: On the epistemic implications of secret Internet technologies. *Episteme, 10*(2), 117–134.

Mitchell, T. (1997). *Machine learning.* McGraw-Hill.

Mittelstadt, B., Allo, P., Taddeo, M., Wachter, S., & Floridi, L. (2016). The Ethics of Algorithms: Mapping the Debate (November 1, 2016). *Big Data & Society, 3*(2), 1–19.

Morek, R. (2006). Regulatory framework for online dispute resolution: A critical view. *The University of Toledo Law Review, 38*, 163.

Mowshowitz, A., & Kawaguchi, A. (2002). Bias on the web. *Communications of the ACM, 45*(9), 56–60.

Newman, D. T., Fast, N., & Harmon, D. (2016). *When eliminating Bias isn't fair: Decision making algorithms and organizational justice.* Presented at the Society for Business Ethics in Anaheim, CA.

Neyland, D. (2016). Bearing account-able witness to the ethical algorithmic system. *Science, Technology and Human Values, 41*(1), 50–76.

Newell, S., & Marabelli, M. (2015). Strategic opportunities (and challenges) of algorithmic decision making: A call for action on the long-term societal effects of 'datification'. *The Journal of Strategic Information Systems, 24*(1), 3–14.

Naik, G., & Bhide, S. S. (2014). Will the future of knowledge work automation transform personalized medicine? *Applied & Translational Genomics, Inaugural Issue, 3*(3), 50–53.

Noble, S. U. (2018). *Algorithms of oppression: How search engines reinforce racism.* NYU Press.

Parsons, C. (2015). Beyond privacy: Articulating the broader harms of pervasive mass surveillance. *Media and Communication, 3*(3), 1–11.

Raymond, A. (2015). The Dilemma of Private Justice Systems: Big Data Sources, the Cloud and Predictive Analytics. *Northwestern Journal of International Law & Business, 35* (1A); Academy of Legal Studies in Business National Proceedings, 45 (2014).

Romei, A., & Ruggieri, S. (2014). A multidisciplinary survey on discrimination analysis. *The Knowledge Engineering Review, 29*(5), 582–638.

Rubel, A., & Jones, K. M. L. (2014). *Student privacy in learning analytics: An information ethics perspective.* SSRN Scholarly Paper,Social Science Research Network.

Russell, S., & Norvig, P. (2009). *Artificial intelligence: a modern approach* (3rd ed.). Prentice Hall Press.

Samoili, S., Cobo, M. L., Gomez, E., De Prato, G., Martinez-Plumed, F., & Delipetrev, B. (2020). *AI Watch. defining artificial intelligence. Towards an operational definition and taxonomy of artificial intelligence* (No. JRC118163). Joint Research Centre (Seville site).

Sandvig, C., Hamilton, K., Karahalios, K. &. Langbort, C. (2016). When the algorithm itself is a racist: Diagnosing ethical harm in the basic components of software. *International Journal of Communication, 10*, 4972–4990.

Shannon, C. E. & Weaver, W. (1998). *The mathematical theory of communication.* University of Illinois Press.

Schreurs, W., Hildebrandt, M., Kindt, E. & Vanfleteren, M. (2008). Cogitas, ergo sum. The role of data protection law and nondiscrimination law in group profiling in the private sector. In Hildebrandt, M., & Gutwirth, S. (Eds.), *Profiling the European citizen: Cross-disciplinary perspectives* (pp. 241–270). Springer.

Schermer, B. (2013). Risks of profiling and the limits of data protection law. In Custers, B., Calders, T., Schermer, B., et al. (Eds.), *Discrimination and privacy in the information society* (pp. 137–152). Springer.

Schermer, B. W. (2011). The limits of privacy in automated profiling and data mining. *Computer Law & Security Review, 27*(1), 45–52.

Sculley, D., Holt, G., Golovin, D., Davydov, E., Phillips, T., Ebner, D., Chaudhary, V. & Young, M. (2014). Machine learning: The high interest credit card of technical debt. *NIPS 2014 Workshop.*

Shackelford, S. J., & Raymond, A. H. (2014). Building the virtual courthouse: Ethical considerations for design, implementation, and regulation in the world of Odr. *Wisconsin Law Review*, (3), 615–657.

Stark, M., & Fins, J. J. (2013). Engineering medical decisions. *Cambridge Quarterly of Healthcare Ethics, 22*(4), 373–381.

Sullins, J. P. (2006). When is a robot a moral agent? *International Review of information ethics, 6*(12), 23–30.

Sunstein, C. (2007). *Republic.com 2.0.* Princeton University Press.

Turilli, M. (2007). Ethical protocols design. *Ethics and Information Technology, 8*, 253–262.

Turilli, M., & Floridi, L. (2009). The ethics of information transparency. *Ethics and Information Technology, 11*(2), 105–112.

Tutt, A. (2016). *An FDA for algorithms.* SSRN Scholarly Paper, *Social Science Research Network.* Available at: http://papers.ssrn.com/abstract=2747994 (accessed 22 July 2020).

Van, W. L., & Royakkers, L. (2004) Ethical issues in web data mining. *Ethics and Information Technology, 6*(2), 129–140.

Vapnik, V. N. (2000). *The nature of statistical learning theory.* Springer.

Wiener, N. (1988). *The human use of human beings: Cybernetics and society.* Da Capo Press.

Zarsky, T. (2016). The trouble with algorithmic decisions an analytic road map to examine efficiency and fairness in automated and opaque decision making. *Science, Technology & Human Values, 41*(1), 118–132.

2 Overview of AI taxonomy and technological methods

This chapter will depict the multilayered conceptualization of AI and its various technological methods. The literature on AI is gaining more scholarly attraction, due to the detrimental implications of AI usage in the public sphere. A more oriented approach will be suggested to enhance the confrontation of the public sector with these technological obstacles and maximize the full potential of Public AI. Broadly, the discourse of AI is not considered relatively new, and yet its definition remains ambiguous among scholars, policy makers, and tech practitioners (Kirsh, 1991; Allen, 1998; Simon et al. , 2000; Brachman, 2006; Nilsson, 2009; Bhatnagar et al., 2017; Monett & Lewis, 2017). Therefore, the term 'AI' has received various interpretations in different human disciplines.

The increasing AI usage in human domains requires a better understanding of the way the scholarly community addresses these complex algorithmic models. The literature highlights a broad public understanding of AI as a transformative agent in human reality:

> Rapid advances in computing power, the increasing availability of data and of new algorithms, has recently led to major breakthroughs in the field of Artificial Intelligence (AI) and let emerge the great potential of this 'new set of technologies' to transform our societies and economic systems, becoming one of the most important technologies of the century for citizens, industry and governments alike.
>
> (Misuraca & van Noordt, 2020, p. 7)

In addition, scholars offer a broad AI definition as: '[t]he study of the computations that make it possible to perceive, reason, and act' (Patrick, 1992). AI is also defined as the automation processes of intelligent

DOI: 10.4324/9781003106678-3

conduct (Luger & Stubblefield, 2008). The Council of Europe offers another definition of AI as

> [s]et of sciences, theories and techniques whose purpose is to reproduce by a machine the cognitive abilities of a human being. Current developments aim, for instance, to be able to entrust a machine with complex tasks previously delegated to a human.
> (Council of Europe Commissioner for Human Rights, 2019)

Additionally, as further articulated by McCarthy (1988):

> AI is concerned with methods of achieving goals in situations in which the information available has a certain complex character. The methods that have to be used are related to the problem presented by the situation and are similar whether the problem solver is human, a Martian, or a computer program.

Following the definitions above, as suggested by the review of AI Watch (Samoili, Cobo, Gomez, De Prato, Martinez-Plumed, & Delipetrev, 2020), most AI definitions highlight the machine's replication of actions that require intelligence (Bellman, 1978), and others attempt to oversimplify the notion of intelligence and define rational AI (Russell & Norvig, 2010; HLEG, 2019). Nonetheless, other studies suggest notable aspects associated with AI, such as the perception of the environment, inputs collection and interpretation, autonomous decision making, and goals achievement.

Between AI and intelligence

Most scholars associate AI with the ambiguous notion of intelligence, and others refer to AI within the notion of rationality. (Gardner, 1987, 1983; Poole, Mackworth, & Goebel, 1998) Meaning, one's '[a]bility to choose the best action to take in order to achieve a certain goal, given certain criteria to be optimized and the available resources.' (High-Level Expert Group on Artificial Intelligence, 2019, p. 1). Others define AI as the machine's attempt to act and think humanly (Yavuz, 2019). Notwithstanding, the common understanding of AI intelligence is far from being unanimous and is highly controversial among scholars. Nonetheless, it is well known that human beings differ from machines in their mental ability and intelligence, and in this regard, AI is the technical attempt to mimic this ability in computer systems and devices.

Notably, the scholar Pei Wang mentions that since AI is not meant to duplicate human intelligence, and varied AI systems are different from humans in numerous aspects, which should be noted (Wang, 1995a). Following the above, it is important to understand both the similarities and limitations of AI-driven intelligence, compared to humans.

This semantic complexity can be explained through the constant modifications in AI usage and goals. In the 1940s, with the invention of the computer, people started to understand that computerized systems can execute an abundant number of tasks that require human intelligence and not just numerical calculations. One of the common definitions for intelligence refers to 'the ability to solve hard problems' (Minsky, 1985, 2006). More specifically, intelligence may be defined as the ability to understand and apply knowledge or the ability to exercise reason. In this regard, it is important to mention the vague nature of intelligence, which combines several cognitive functions, including attention, memory, language, perception, and planning (Colom et al., 2010). Additional intelligence definitions suggest its ability to '[g]ather, to collect, to assemble or to choose, and to form an impression, thus leading one to finally understand, perceive, or know' (De Spiegeleire, Maas & Sweijs, 2017, p. 26). Another interesting interpretation divides intelligence into two types; natural and mechanical (Newell, 1992). While natural intelligence is based upon the five human senses, mechanical intelligence includes artificial sensors controlled by a mechano-neural network, and the data is collected by machines through simulating nature (Patterson, 1990).

An additional definition was offered by Wang (1995b, p. 1): 'Intelligence is the capacity of an information-processing system to adapt to its environment while operating with insufficient knowledge and resources'. From a different perspective, several scholars offer a social perceptive of AI, as a '[t]echno social system', in which the technical features of AI are inherently connected to its social aspects. In this regard, social values impact the way society perceives and use AI technologies. (Hagerty & Rubinov, 2019; Beer, 2009).

Following the above, Wang overviews different ways to define AI, which observes the similarity to human intelligence in terms of structure, behaviour, function and principle. For instance, structure-wise, some may suggest that since intelligence derives from the human brain, AI can be created through a brain-like structure, as seen in previous developments such as Connection Machine and Artificial Neural Networks (Hillis, 1989; Smolensky, 1988; Newell, 1994). However, it is important to mention that with the high complexity of the human brain and its significant difference from computing devices, it is still doubtful whether AI can be the outcome of a brain structure. In comparison, others suggest defining AI by behaviour, as seen in the Turing Test. A proponent of this approach is the scholar Allen Newell who presented the Soar project as an AI system and a cognitive model, where the computer-produced outputs are compared to a psychological dataset, created by humans.

National security perspective of AI taxonomy

Due to the increasing AI-driven applications in the military sphere, one may mention notable uses of AI for national security purposes, which highlight its varied operational functionalities. For instance, the 2019 US National Defense Authorization Act offers a valuable starting point. It defines AI as including

> (1) any artificial system that performs tasks under varying and unpredictable circumstances without significant human oversight, or that can learn from experience and improve performance when exposed to data sets; (2) artificial systems developed in computer software, physical hardware, or other context that solves tasks requiring human-like perception, cognition, planning, learning, communication, or physical action; (3) artificial systems designed to think or act like a human, including cognitive architectures and neural networks; (4) a set of techniques, including machine learning, that is designed to approximate a cognitive task; and (5) artificial systems designed to act rationally, including an intelligent software agent or embodied robot that achieves goals using perception, planning, reasoning, learning, communicating, decision making, and acting.
>
> (Congress, US, 2018, p. 330)

In addition, the US Army Sciences Board highlighted the decision making aspect of AI, that it 'can incorporate abstraction and interpretation into information processing and make decisions at a level of sophistication that would be considered intelligent in humans' (Brownstein and et al., 1984), or as articulated by the US Defense Science Board, which depicts AI as '[t]he capability of computer systems to perform tasks that normally require human intelligence (e.g., perception, conversation, decision-making)' (Fields, 2016).

The more streamlined definition offered by the European Commission's Communication on AI refers to AI as '[s]ystems that display intelligent behaviour by analysing their environment and taking actions—with some degree of autonomy—to achieve specific goals'. It posits that AI-based systems 'can be purely software-based, acting in the virtual world (e.g. voice assistants, image analysis software, search engines, speech and face recognition systems) or AI can be embedded in hardware devices (e.g. advanced robots, autonomous cars, drones or Internet of Things applications)' (EU Commission, 2019, p. 1).

Notably, whereas the American definition is more specific and detailed when it comes to the technology, the European definition addresses AI as a holistic concept while putting the emphasis on the intelligent behaviour of machines and their varying level of autonomy. Both definitions are helpful in understanding the broad spectrum of technologies with potentially substantial outcomes for security and economy.

As previously mentioned, AI technologies consist of a computer program that can learn, act, and adapt in a relatively similar manner to humans. However, it consists of computational information processing and not biological. Broadly, AI's power is based upon the exponential growth in computational processing and storage and advances in robotics. Some may argue that the future development of quantum computing may lead to a significant enchantment of AI capabilities (Wadhwa; Manheim, & Kaplan, 2018).

Within this security framework, it is important to mention the emergence of autonomous weapons. A common definition suggests the following: 'Any weapon system with autonomy in its critical functions— that is, a weapon system that can select (search for, detect, identify, track or select) and attack (use force against, neutralize, damage or destroy) targets without human intervention' (Davison, 2018, p. 5). This term has led to a broad legal discussion regarding the ability to employ lethal autonomous weapons systems, using AI technologies, to identify targets and destroy them without human control (Sayler, 2019). Important to mention, this dystopian vision of machine autonomy is still not feasible on the battlefield and is mainly controlled by humans. However, one may note the increasing deployment of drones for these purposes (Lubell & Derejko, 2013; Lewis, 2011; Tullington, 1984).

AI techniques and evolutionary stages (ANI, AGI, & ASI)

Following the varied AI definitions, it is important to mention that AI encompasses an abundant number of algorithms and technological techniques. For instance, machine learning is considered an advanced form of AI, which relies upon statistical methods and learning from data. Notably, machine learning can be human-trained, 'supervised', or self-trained, 'unsupervised'. In this regard, a prevalent form of machine learning is deep learning, which uses artificial neural networks that attempt to mimic the structure of the human brain. Their sophisticated structure enables them to retrieve patterns in non-structured data (Kavukcuoglu, Graepel, & Hassabis, 2016; Kurzweil, 2005).

These technological developments in the last few decades have led to the flourishing of AI research and differences in AI taxonomy, with emphasis on deep learning, highlighting a more sophisticated approach to AI usage (Silver et al., 2016; LeCun, Bengio, & Hinton, 2015). In addition, new terms emerged, such as 'Human-level AI' (McCarthy, 2007; Nilsson, 2005; Minsky, Singh, & Sloman, 2004) and 'Artificial General Intelligence' (AGI) (Goertzel & Pennachin, 2007; Wang & Goertzel, 2007). It is important to mention that several scholars suggest another term 'Strong AI' (Searle, 1980).

In this regard, the literature further suggests a notable AI distinction to its three granular and consecutive evolutionary steps: narrow artificial intelligence, general artificial intelligence, and superior artificial Intelligence. An additional essential AI taxonomy relates to super intelligence and to AI's alleged ability to outweigh human intelligence. Currently, it is suggested that AI is still in its preliminary stage of narrow intelligence, where certain computer programs outperform human individuals in specific tasks such as reading X-rays and writing legal contracts. Notable examples include, but are not limited to, IBM's Deep Blue (Chess), Google's AlphaGo (go), and High-Frequency Trading Algorithms. In the future, some estimate that AI will transform into general intelligence, and its capabilities will exceed the current problem solving, and even lead to super intelligence, when AI will outperform the smartest human individuals (Manheim, & Kaplan, 2018; Kurzweil, 2005; Licklider, 1960).

Algorithmic bias

Within the AI characteristics, it is important to mention the prevalent term of 'Algorithmic Bias', in which algorithms exhibit existing human biases in the programming and training process. Therefore, biased sampling or any other algorithmic limitation may produce imperfect data, and AI-driven decisions may exacerbate human biases, and pose significant threats to equality and liberal values. There are varied ways in which AI can reflect bias. For instance, if the historical data in crime statistics is racially biased against certain groups, AI-driven decision making is more likely to amplify the bias in each legal process (Manheim, & Kaplan; Sherman, 2018). In addition to data training, the data sets which are used to produce AI models may also be biased. For instance, input data may be biased since it is generated by humans or sensors that are still designed by humans (Shorey & Howard, 2016). Following the above, one may mention several examples of algorithmic bias in AI technologies. The example of Microsoft's Tay, a teen-talking AI chatbot, exhibits this bias, as the chatbot made discriminatory tweets within a short time, due to her machine learning capabilities, even when not designed to mimic human malicious intent (Kleeman, 2016). In addition, facial recognition software often does not detect darker skin or classifies black individuals as animals (Dougherty, 2015). Moreover, the results of Google's search algorithm and Amazon's recruiting tool reflected occupational gender stereotypes (Calo, 2017; Holley, 2018). However, as articulated by Manheim and Kaplan (2018):

> It is impossible to strip bias from human beings, but it may be possible to remove bias from AI with the proper governance of data input. Do we want AI to reflect the stereotypes and discrimination prevalent in society today, or do we want AI to reflect a better society where all people are treated as equal? Timnit Gebru, co-founder of the Black in AI event, advocates that diversity is urgently needed in AI. This means more than a variety of people working on technical solutions and includes diversity in data sets and conversations about law and ethics [...] Governance over data input is thus necessary to ensure it is vast, varied, and accurate.

Due to the increasing impact of algorithmic decision making processes on varied human aspects, such as finance, healthcare, hiring, policy making and education, it is of paramount importance to discuss the societal and ethical implications of possible discriminatory patterns of such technologies, such as gender, health, or ethnicity (Hajian, Bonchi, & Castillo, 2016). The next chapters will elaborate upon these challenges posited by AI technologies.

References

Allen, J. F. (1998). AI growing up: the changes and opportunities. *AI Magazine*, *19*(4), 13–23.

Bhatnagar, S., Alexandrova, A., Avin, S., Cave, S., Cheke, L., Crosby, M., ... & Hernández-Orallo, J. (2017, November). Mapping intelligence: Requirements and possibilities. In *3rd Conference on" Philosophy and Theory of Artificial Intelligence* (pp. 117–135). Springer, Cham.

Beer, D. (2009). Power through the algorithm? Participatory web cultures and the technological unconscious. *New Media & Society*, *11*(6), 985–1002.

Bellman, R. (1978). *An introduction to artificial intelligence: can computers think?* Thomson Course Technology.

Brachman, R. J. (2006). (AA)AI—more than the sum of its parts, 2005 AAAI Presidential Address. *AI Magazine*, *27*(4), 19–34.

Brownstein, B. J., Reidy, J. J., & Renner, G. F. (1984). *Technological assessment of future battlefield robotic applications.* Defense Technical Information Center.

Calo, R. (2017). Artificial intelligence policy: a primer and roadmap. *UCDL Review*, *51*, 399, 411–12.

Colom, R., Karama, S., Jung, R. E., & Haier, R. J. (2010). Human intelligence and networks. *Dialogues in Clinical Neuroscience*, *12*, 489–501.

Congress, US (2018). One Hundred Fifteenth Congress of the United States of America. *At the second session. Begun and held at the City of Washington on Wednesday, the third day of January, two thousand and eighteen. An Act. To authorize appropriations for fiscal year 2019 for military activities of the Department of Defense, for military construction, and for defense activities of the Department of Energy, to prescribe military personnel strengths for such fiscal year, and for other purposes. HR, 5515.*

Council of Europe Commissioner for Human Rights (2019). *Unboxing artificial intelligence: 10 steps to protect human rights.* Council of Europe.

Davison, N. (2018). *A legal perspective: Autonomous weapon systems under international humanitarian law.* International Committee of the Red Cross.

De Spiegeleire, S., Maas, M., & Sweijs, T. (2017). *Artificial intelligence and the future of defense: strategic implications for small-and medium-sized force providers.* The Hague Centre for Strategic Studies.

Dougherty, C. (2015). Google photos mistakenly labels black people 'gorillas'. *The New York Times, 1.*

EU Commission. (2019). *A definition of AI: main capabilities and disciplines.* Independent High-Level Expert Group on Artificial Intelligence set by the European Commission.

Fields, C. (2016). *Report of the defense science board summer study on autonomy.* Office of the Under Secretary of Defense for Acquisition, Technology and Logistics.

Gardner, H. (1983). *Frames of mind; The theory of multiple intelligences.* Basic Books.

Gardner, H. (1987). *The mind's new science: A history of the cognitive revolution.* Basic Books.

Goertzel, B., & Pennachin, C. (Eds.). (2007). *Artificial General Intelligence.* Springer.

Hagerty, A., & Rubinov, I. (2019). *Global AI ethics: a review of the social impacts and ethical implications of artificial intelligence.* Cornell University. Available at: arxiv.org/abs/1907.07892v1

Hajian, S., Bonchi, F., & Castillo, C. (2016, August). Algorithmic bias: From discrimination discovery to fairness-aware data mining. In *Proceedings of the 22nd ACM SIGKDD international conference on knowledge discovery and data mining* (pp. 2125–2126).

Hillis, W. D. (1989). *The connection machine*. MIT press.

HLEG, A. (2019). High-level expert group on artificial intelligence. *Ethics Guidelines for Trustworthy AI*. European Commission.

Holley, P. (2018), This Patent Shows Amazon May Seek to Create a 'Database of Suspicious Persons' Using Facial-Recognition Technology. *Washingtonpost. com,* December 12, 2020, from: www.washingtonpost.com/technology/2018/ 12/13/this-patent-shows-amazon-may-seek-create-database-suspicious-persons-using-facial-recognition-technology.

Kavukcuoglu, K., Graepel, T., & Hassabis, D. (2016). Mastering the game of Go with deep neural networks and tree search. *Nature, 529*, 484–489.

Kleeman, S. (2016). Here Are the Microsoft Twitter Bot's Craziest Racist Rants. *Gizmodo. Available at: https://gizmodo.com/here-are-the-microsoft-twitter-bot-s-craziest-racist-ra-1766820160.*

Kirsh, D. (1991). Foundations of AI: the big issues. *Artificial Intelligence, 47*, 3–30.

Kurzweil, R. (2005). *The singularity is near: When humans transcend biology.* Penguin.

LeCun, Y., Bengio, Y., & Hinton, G. (2015). Deep Learning. *Nature, 521*, 436–444.

Lewis, M. W. (2011). Drones and the Boundaries of the Battlefield. *Texas. International Law Journal, 47*, 293.

Licklider, J. C. (1960). Man-computer symbiosis. *IRE Transactions on Human Factors in Electronics*, (1), 4–11.

Lubell, N., & Derejko, N. (2013). A global battlefield? Drones and the geographical scope of armed conflict. *Journal of International Criminal Justice, 11*(1), 65–88.

Luger, G. F., & Stubblefield, W. A. (2008). *Artificial intelligence: structures and strategies for complex problem solving.* Pearson education.

Manheim, K. M., & Kaplan, L. (2018). Artificial intelligence: risks to privacy and democracy. *Yale Journal of Law & Technology, 21*, 177–188

McCarthy, J. (2007). From here to human-level AI. *Artificial Intelligence, 171*, 1174–1182.

McCarthy, J. (1988). Mathematical logic in artificial intelligence. *Dædalus, 117*(1), 297–311.

McCarthy, J. (2007). *What is Artificial Intelligence.* Stanford University.

Minsky, M. (1985). *The society of mind.* Simon and Schuster.

Minsky, M. (2006). The emotion machine: Commonsense thinking. *Artificial Intelligence, and the Future of the Human Mind. Simon & Schuster.*

Minsky, M., Singh, P., & Sloman, A. (2004). The St. Thomas common sense symposium: designing architectures for human-level intelligence. *AI Magazine, 25*(2), 113–124.

Misuraca, G., & van Noordt, C. (2020). AI Watch-Artificial Intelligence in public services: Overview of the use and impact of AI in public services in the EU. *JRC Working Papers*, (JRC120399).

Monett, D., & Lewis, C. W. (2017, November). Getting clarity by defining artificial intelligence—a survey. In *3rd conference on" philosophy and theory of artificial intelligence* (pp. 212–214). Springer, Cham.

Newell, A. (1992). SOAR as a unified theory of cognition: Issues and explanations. *Behavioural and Brain Sciences, 15*(3), 464–492.

Newell, A. (1994). *Unified theories of cognition.* Harvard University Press.

Nilsson, N. J. (2005). Human-level artificial intelligence? Be serious! *AI Magazine, 26*(4), 68–75.

Nilsson, N. J. (2009). *The quest for artificial intelligence.* Cambridge University Press.

Patrick, W. H. (1992). *Artificial intelligence* (3rd ed.). Addison-Wesley Publishing Company.

Patterson, D. W. (1990). *Introduction to artificial intelligence and expert systems.* Prentice-Hall of India.

Poole, D., Mackworth, A., & Goebel, R. (1998). *Computational intelligence: a logical approach.* Oxford University Press.

Russell S. & Norvig P.(2010). *Artificial intelligence - A modern approach.* Third International Edition. Pearson Education.

Samoili, S., Cobo, M. L., Gomez, E., De Prato, G., Martinez-Plumed, F., & Delipetrev, B. (2020). *AI Watch. Defining Artificial Intelligence. Towards an operational definition and taxonomy of artificial intelligence* (No. JRC118163). Joint Research Centre (Seville site).

Sayler, K. M. (2019). *Defense Primer: US Policy on Lethal Autonomous Weapon Systems.* Congressional Research Service.

Searle, J. (1980). Minds, brains, and programs. *The Behavioural and Brain Sciences, 3*, 417–424.

Silver, D., Huang, A., Maddison, C. J., Guez, A., Sifre, L., van den Driessche, G., Schrittwieser, J., Antonoglou, I., Panneershelvam, V., Lanctot, M., Dieleman, S., Grewe, D., Nham, J., Kalchbrenner, N., Sutskever, I., Lillicrap, T., Leach, M., Kavukcuoglu, K., Graepel, T., & Hassabis, D. (2016). Mastering the game of Go with deep neural networks and tree search. *Nature, 529*(7587), 484–849.

Simon, H. A., Bibel, W., Bundy, A., Berliner, H., Feigenbaum, E. A., Buchanan, B. G., ... & McCarthy, J. (2000). AI's greatest trends and controversies. *IEEE Intelligent Systems and Their Applications, 15*(1), 8–17.

Sherman, J. (2018). *AI And Machine Learning Bias Has Dangerous Implications*, OPEN-SOURCE. Retrieved December 13, 2020, from: https://opensource. com/article/18/1/how-open-source-can-fight-algorithmic-bias

Shorey, S., & Howard, P. (2016). Automation, big data and politics: a research review. *International Journal of Communication, 10*, 5032–5055.

Smolensky, P. (1988). On the proper treatment of connectionism. *Behavioural and Brain Sciences, 11*(1), 1–23.

Tullington, B. J. (1984). *Proceedings of the Army Conference on Application of Artificial Intelligence to Battlefield Information Management Held at White Oak, Maryland on April 20, 21, and 22, 1983.* Defense Technical Information Center.

US National Defense. (2018). Authorization Act for Fiscal Year 2019.

Wang, P. (1995a). *Non-axiomatic reasoning system: Exploring the essence of intelligence.* Bloomington, IN: Indiana University.

Wang, P. (1995b). *On the working definition of intelligence. Center for Research on Concepts and Cognition.* Indiana University.

Wang, P., & Goertzel, B. (2007). Introduction: Aspects of artificial general intelligence. In Goertzel, B., and Wang, P., (Eds.), *Advance of Artificial General Intelligence* (pp. 1–16). IOS Press.

Wadhwa, V. (2018). Quantum Computers May Be More of an Imminent Threat than AI. *Washington Post, 5.* Retrieved December 12, 2020, from: www.washing-tonpost.com/news/innovations/wp/2018/02/05/quantum-computers-may-be-more-of-an-imminent-threat-than-ai.

Yavuz, C. (2019). *Machine Bias: Artificial Intelligence and Discrimination.* Available at SSRN 3439702.

3   Developing ethical framework for ethical responsible innovation

This chapter presents existing approaches for responsible innovation governance. Responsible innovation (RI) often refers to the institutional structure for innovation that can lead to well accepted technological advances. In this chapter, we aim to bring ethical aspects of RI governance to society, so that they can facilitate decision-making by stakeholders. Moral philosophers and applied ethicists often suggest using a 'deontological' approach to moral norms as a starting point, while others suggest a 'teleological' (consequentialist) ethical approach concerned with whether an action or decision leading to an outcome is good or bad to the society as a whole (Gomila & Amengual, 2009). It is argued that for AI as machines with a more limited degree of autonomy, a rule-based approach may be sufficient, or even endorsed. Since these theories are based on explicit rules and norms external to the real world in which they are to be applied, they have limited practical value. In other words, these theories are, from a philosophical perspective, unable to fully explicate the complexities of moral considerations as these complexities are experienced in the world.

This chapter suggests that deontological and utilitarian ethics cannot fully address the challenges that AI technological innovations pose to contemporary ethics, which require a more flexible, context-dependent, and case-based approach to morality. For that, we introduce the conceptual laying of prudential rationality ethics. This ethical notion brings space into the analysis of the application of machine learning, cognitive computing, algorithmic, and data governance.

Governance of responsible innovation

Technological innovations have often been invoked by ethical impact and concerns. For that, the responsibilities of scientists to produce reliable knowledge often reach beyond those associated with their professional

DOI: 10.4324/9781003106678-4

expertise. (Douglas, 2003; Mitcham, 2003; Steneck, 2006). The negotiation of responsibility between practicing scientists, innovators, and society remains a contested research field. Studies of research in Science and Technology (STS) call for broadening the normative grounding of innovation responsibility to include social and political aspects and to address the implications of the overlapping boundaries between existing scientific regulatory institutional frameworks (Callon et al., 2009). According to Callon et al. (2009) science and technology often constitute 'continuous movement toward a greater and greater level of attachments of things and people at an ever expanding scale and at an ever increasing degree of intimacy' (Latour, 2008, p. 4). These observations lead to associate responsibility in governance with risk-management by framing responsibility in terms of accountability or liability (Pellizzoni, 2004; Grinbaum & Groves, 2013).

The growing interest in and importance of responsible research and innovation (RRI) and the need to extend the scope of governance processes beyond formal risk assessment, have lead René Von Schomberg (2013, p. 63), the European Commission's Directorate General for Research and Innovation, to define in advance responsible innovation research: 'Responsible research and innovation is a transparent, interactive process by which societal actors and innovators become mutually responsive to each other with a view to the (ethical) acceptability, sustainability and societal desirability of the innovation process and its marketable products (in order to allow a proper embedding of scientific and technological advances in our society)' (Von Schomberg, 2013, p. 63).

Von Schomberg's (2013) conceptualization addresses the belief that public concerns about technological innovation cannot be limited to risk assessment, but rather encompass a range of ethical principles and approaches that should be incorporated into ethical innovation governance (Macnaghten & Szerszynski, 2013; Stilgoe et al., 2013; Smith et al., 2005; Stirling, 2008; Morlacchi & Martin, 2009; Wilsdon & Willis, 2004; Blok & Lemmens, 2015; Nathan, 2015). Von Schomberg (2011) has developed a RI framework:

A transparent, interactive process by which societal actors and innovators become mutually responsive to each other with a view to the (ethical) acceptability, sustainability and societal desirability of the innovation process and its marketable products (in order to allow a proper embedding of scientific and technological advances in our society).

(Von Schomberg, 2011, p. 740)

Drawing on this process-based approach to RI governance, Stilgoe et al. (2013) offer a four-dimensional ethical framework that includes anticipation, reflexivity, inclusion, and responsiveness.

The dimension of anticipation refers to the scholarly critiques of the limitations of top-down risk-based models of governance to address the social, ethical, and political concerns associated with technological innovations (Jasanoff, 2003; Henwood & Pidgeon, 2013). It is suggested that anticipation can improve resilience in the context of risk-assessment (Brown et al., 2000; Fortun, 2001; Brown & Michael, 2003; Martin, 2010; Borup et al., 2006; Selin, 2007). Fortun (2005) uses the term 'an ethics of promising' to demonstrate an urgent need for critical reflection on overly-optimistic scientific expectations, which tend to predict and also shape desirable futures (Te Kulve & Rip, 2011). A process of anticipation is required to balance between the value of prediction, which tends to reify particular futures, and public engagement to raise concerns and complexities embedded in the uncertainties of science and the dynamics of social affairs. (Barben et al., 2008; Wilsdon & Willis, 2004).

Methods applied to underscore the implications of anticipatory processes require reflection on the potentially unintended impacts of technological innovations (Selin, 2011; Robinson, 2009; Grin & Grunwald, 2000). The use of scenarios is viewed as an effective method in establishing a modified set of structured conceptual systems of equally plausible future contexts, often presented as narrative descriptions, targeted to provide inputs for the anticipatory process (Selin, 2011). While social scientists have applied the scenario method extensively, the use of scenarios in RI research must manage institutional and cultural resistance to anticipation (Stilgoe et al. 2013; Guston, 2012). An additional method applied to the anticipatory process is by drawing on science fiction to democratise reflection of possible harms and benefits of future technologies (Miller & Bennett, 2008; Borup et al., 2006; Selin, 2011).

The dimension of reflectivity in building the RI framework raises the need for institutional reflexivity in governance. Reflexivity allows engaging in a process of continuous learning. Reflexivity is closely linked to the concept of experience by being aware of one's own activities, commitments, and assumptions (Wynne, 2011; Schuurbiers, 2011; Schuurbiers & Fisher, 2009). As such, reflexivity is a strategy to explore a set of values, assumptions, prejudices, and habitual actions, to strive to understand the shaping of our surroundings, and begin to critically take circumstances and relationships into the decision-making process. Building reflectivity needs to incorporate 'reflections of natural scientists on the socio-ethical context of their work' (Schuurbiers, 2011,

p. 769). Therefore, tools such as codes of conduct, moratoriums, and the adoption of standards may complement the process of reflexivity by drawing connections between external value systems and scientific practice (Von Schomberg, 2013). The challenge lies in the fact that building actors' and institutions' reflexivity means rethinking prevailing conceptions about the practice of science and innovation and the roles of scientists and institutions in developing responsibility and accountability within RI decisions. Methods applied to enhance reflectivity include action research and adaptive management.

The dimension of Inclusion Considerable research exists on the inclusion of public voices in the governance of science and innovation as a strategy of building legitimacy, (Macnaghten & Chilvers, 2014; Lövbrand et al., 2011; Dryzek, 2011; Felt& Wynne , 2007; Irwin, 2006; Stirling, 2006; Wilsdon & Willis, 2004; Von Hippel, 1976, 2005). This was taken in the form of deliberative forums and focus groups on issues of technological innovations involving the inclusion of the public interest and the balancing between serving narrow stakeholders' interests and the broader public interest. The understanding of the complexity of regulatory mechanism applied in the scientific innovation domain has led to incorporate an ethical and normative perspective that considers the pivotal role of public engagement in decision-making processes (Chilvers, 2013; Goodin & Dryzek, 2006).

However, the practice of inclusive governance and its impact on policymaking has attracted criticism, and is often accused that 'the technical is political, the political should be democratic and the democratic should be participatory' (Moore, 2010, p. 793). In addition, Irwin et al. (2013) argue that '[t]he (often implicit) evocation of the highest principles that engagement might ideally fulfill can make it difficult to acknowledge and pay serious attention to the varieties of engagement that are very much less than perfect but still somehow "good"' (p. 120).

One way to address these limitations is to use public dialogue as a public good in itself (Chilvers, 2013). Scholarly attention has been drawn to the introduction of participatory approaches activities as a learning process to frame the effects of innovations within dialogue processes (Rowe & Frewer, 2000, 2005; Stirling, 2008). Callon et al. (2009, p. 160) offer three criteria to evaluate the effectiveness of public dialogue in RI including *intensity*, that is, how early members of the public are consulted and how much care is given to the composition of the discussion group; *openness* as to how diverse the group is and who is represented; and *quality*, that refers to the gravity and continuity of the discussion.

The dimension of responsiveness embedded in RI framework assures that those technological innovations proven effective are made available to the population. Research outputs certainly should be made widely available in a way that maximizes their value and usefulness. According to Pellizzoni, responsiveness is defined as 'an encompassing yet substantially neglected dimension of responsibility' (Pellizzoni, 2004, p. 557). Pellizzoni (2004) combines the two aspects associated with the responsiveness that is, to react and to answer. As such responsiveness involves responding to new knowledge as this emerges and to emerging concerns, views, and norms as well as providing institutional practices to address these concerns (Stilgoe et al., 2013).

Responsiveness in RI demands that institutions and technology assessment and foresight processes may be improved by accessibility and availability of stakeholders' concerns and needs. Awareness of and appreciation for ethical challenges that arise from the distinctive features of scientific innovation must be discussed with diverse populations when reviewing the RI governance framework (Von Schomberg, 2013). For example, citizens may be able to identify harmful effects that in technological innovation settings are difficult to estimate and are likely to evolve as the innovation implementation process unfolds. Stakeholders may also discuss the need to consider issues of security, including the safety of data collectors, and research participants. As such, responsiveness allows acknowledging that standard scientific approaches may need to be adapted and circumscribed by societal challenges and the realities of applying technological innovations (Von Schomberg, 2013; Stirling, 2006).

In addition, being receptive also means that RI resists static representations of individuals' identities and social locations, but rather gives proper attention and investigation to diversity. The use of social groupings no doubt helps to simplify the complexities of interests and needs of society, especially for the purpose of RI decision-making and policy. However, their use is often premised on assumptions that such groups have similar predispositions, concerns, and needs that require similar approaches and policy solutions, when actually no homogenous social group or unitary perception of society exists (Schot & Geels, 2008). Responsiveness must address explicitly the tensions and governance mechanisms within processes of technological innovation governance.

Various tools to enhance the aspects of institutional responsiveness on decision acceptance and the legitimacy of technological innovations include: a deliberative science policy culture, engagement of citizens in reflexive learning and decision-making processes, distinguishing between citizens' interests and their concerns, creativity and innovation, interdisciplinarity, risk-evaluation and management, ethical entrepreneurship leadership and commitment to the public interest, and commitments to openness and transparency (Macnaghten & Chilvers, 2014).

The framework for responsible innovation presented by Stilgoe et. al (2013) prompts new directions and challenges to go beyond existing regulation and institutional responsibilities of technological innovations. Although recent efforts to build ethical innovation governance attempt to incorporate ethical principles and standards (e.g., Owen et al., 2012; Pavis et al., 2014), the aspects of responsible innovation should be embedded in ethical decision-making frameworks (Nathan, 2015). In the next section, we discuss two main normative ethical decision-making theories, that is, theories for moral and evaluative judgements in ethical decision-making of RI, namely deontology and utilitarianism (consequentialism).

Ethical approaches to responsible innovation: Deontology and Consequentialism

Ethical aspects of RI governance which aim to facilitate a decision maker to face with challenges of technological innovations are usually divided into two subareas. The first, associated with the deontological approach, is represented by Kant's (1785/1988) demand for categorically guided action through reason. Kant's first formulation of the categorical imperative is that it is a principle with which everybody could reasonably agree. Kant's second formulation of the categorical imperative underlines that human beings should always be treated as ends in themselves, not merely instrumentally as a means to an end. The process of ethical reasoning based on rights and duties is designated as deontology (from the Greek *deon* = obligation). Deontological theories hold that actions are based on duties or principles that are intrinsically right or wrong and thus obligatory or forbidden, regardless of the motives for which they are performed or the states of affairs in which they result. A deontological ethics approach includes a fixed set of duties, rules and policies that define actions as ethical.

Within the deontological evaluation, the individual examines the appropriateness or wrongness of the behaviours led by each alternative. He or she must compare each alternative's behavioural practices, based on fixed deontological norms. Pan and Sparks (2012, p. 413) explain that deontological evaluations: '[o]ccur as decision makers employ laws, rules, codes or behavioural norms to each perceived alternative'.

Important to mention, these norms or laws do not have to be highly general nor specific, and the individual examines the degree to which the perceived decision alternative disrupts these socially applicable codes.

The second approach, associated with the utilitarian philosophies of Jeremy Bentham and John Stuart Mill and related to the American school of pragmatism, is a teleological approach that emphasizes the consequences of a given action. For Mill (1871), rights and duties become empty rules if we do not specify what rights these are: '[Social and distributive justice] is involved in the very meaning of Utility, or the Greatest Happiness Principle. That principle is a mere form of words without rational signification, unless one person's happiness, supposed equal in degree (with the proper allowance made for kind), is counted for exactly as much as another's' (Mill, 1871, p. 257).

Consequentialist (utilitarianism is an example) or teleological (from the Greek *telos* or 'end') ethics. Consequentialism ethics allows individuals to choose '[t]he action that will bring the most good to the party the actor deems most important' (Merrill, 2011, p. 11). Following utilitarian ethics, acts are obligatory if they meet the test of the principle of utility, that is if they maximize the happiness of a larger number of people than would alternative courses of action. Actions are then evaluated according to the consequences that they bring about. The saying 'the end justifies the means' presents the essence of a purely consequentialist approach. However, there are two sub-categories of consequentialism. The first is an act-consequentialism in which actions are evaluated based on the consequences of particular acts, and the 'right' action is the one that produces the best consequences in a given situation, compared to its possible alternatives. The second is associated with rule-consequentialism for which actions are evaluated as 'right' if they are in accordance with general rules that produce the best consequences compared to other rules. The consequentialist evaluation suggests that the individual scrutinizes the perceived outcomes from several criteria. He or she examines each alternative for numerous stakeholder groups, its assessed probability, the level of its desirability, and the importance of each stakeholder group (Hunt & Vitell, 1993). Therefore, the theory argues that deontological and teleological evaluations influence one's formation of ethical judgments. As a result, the variance between individuals relies upon their inner choice between deontological and teleological reasoning, as part of their ethical judgements.

Finally, deontological ethics is capable of addressing the considerations that are ignored in consequentialist approaches such as fairness and respect for persons. Deontology can address the importance of treating individuals affected by technological innovations as persons, not merely as means, so that developers, scientists and government actors are responsible for designing a larger role for individuals in RI decision-making. The rights of individuals as persons also have to be respected. However, this raises the practical challenge of whether and how deontology can be applied in RI governance and be aligned with corporate culture and strategy. Instead, a consequentialist approach can be more practically applicable when it comes to balancing potential costs and benefits of technological innovations. Yet, consequentialism is constrained by requirements involving foresight and predictability, which have limitations when associated with technological innovation, so that careful calculation of benefits/costs/risks will not be heuristically helpful for effective innovation governance decision-making.

We, therefore, argue that for the RI governance framework, a rule-based approach may not be sufficient, or even endorsed. Since both deontological and consequentialist ethics are based on explicit rules and norms external to the real world in which they are to be applied, they have limited practical value for ethical decision-making in the domain of technological innovations. In other words, these theories are, from a philosophical perspective, unable to fully explicate the complexities of moral considerations as they are experienced in the world. This chapter suggests that deontological and consequentialist ethics cannot fully address the challenges that technological innovations pose to contemporary ethics, which require a more flexible, context-dependent and case-based approach to morality. For that, we offer to investigate how an ethical framework for technological innovations can be enhanced through critical engagement with prudential ethics.

Prudential ethics and responsible innovation

The wisdom of prudence in politics

A great deal of insight into the idea of prudence in politics can be gleaned from retracing our discussion of the Greek philosophical roots of practical wisdom, particularly the insights gained from Aristotle's and Plato's concepts of virtue (*arete*) that can deepen the incorporation of prudential rationality within ethical egoism. *Arete*, the Greek notion of virtue, denotes excellence in quality that is very much linked to being well-functioning. Aristotle specified three types of human excellence (virtue) including bodily excellence, the excellence of character (moral virtue), and excellence of intelligence (intellectual virtue) (Urmson, 1998).

In Greek ethics, moral virtues or excellence of character refer to characteristics such as courage, temperance and justice. Intellectual virtues encompass wisdom, which includes excellence in theoretical matters and practical wisdom (termed phronesis or prudence). Practical wisdom indicates excellence in practical affairs, and involves the ability to plan one's life well (Urmson, 1998). Viewed in this way, the intellectual virtue of practical wisdom (prudence) cannot be separated from moral virtue (MacIntyre, 1985).

Socrates' philosophy may provide a starting point for an discussion of practical wisdom theory. For Socrates, love, character, harmony, beauty and truth are crucial to wisdom (Birren & Svensson, 2005; Robinson, 1990). Socrates considers wisdom to function as an evolutionary tool in terms of thought and behaviour that contributes to the adaptation and survival of communities. To maintain such function for the long term, Socrates argues that wisdom is needed to exercise the power of both practical and political purposes to bring about well-being. Governments must build expertise and exercise knowledge and good judgment for the sake of society (Osbeck & Robinson, 2005). Similarly, Plato views wisdom as a process of moral deliberation guided by the appropriate balancing of the three parts of the 'soul'—that is, desire, spirit and reason—such capacity is central for wisdom because it maintains harmony between the components of the soul. Wisdom for both Socrates and Plato fits then into moral deliberation and ethical conduct because it provides good judgment leading to good action.

Practical wisdom leads to excellence in ethical deliberation and therefore is a subject of ethics education so that professionals will be able to make sound ethical judgments in practice. According to Aristotle: 'It is impossible to be good in the full sense of the word without practical wisdom or to be a man of practical wisdom without moral excellence or virtue' (Aristotle, 1962, 1144b). For that, Aristotle puts an emphasis on the role of the educator and on the way people can develop moral expertise. It should be noted that viewing virtue as a skill requires different teaching tools that go beyond that which is needed for theoretical wisdom. Teaching practical wisdom and excellence of character should be grounded in everyday practice and cultivated by exemplary practitioners who acquired practical wisdom through years of experience. As articulated by Aristotle (1962, 1140a):

> *We may grasp the nature of prudence [phronesis] if we consider what sort of people we call prudent. Well, it is thought to be the mark of a prudent man to be able to deliberate rightly about what is good and advantageous...But nobody deliberates about things that are*

invariable...So...prudence cannot be a science or art; not science [episteme] because what can be done is a variable (it may be done in different ways, or not done at all), and not art [techne] because action and production are generically different. For production aims at an end other than itself; but this is impossible in the case of action, because the end is merely doing well. What remains, then, is that it is a true state, reasoned, and capable of action with regard to things that are good or bad for man...We consider that this quality belongs to those who understand the management of households or states. [Italics in the original]

The distinctive nature of prudence lies in the fact that it has two forms that one must consider, both of which revolve around the concept of acting in one's self interest: the capacity to gain worldly knowledge, namely objects, and also to deliberate about particular facts, with variables and contingencies. A prudent person, then, is one who is capable of reflecting well on what is good and beneficial for his own interest, taking into consideration elements of contingency and variability which do not permit accurate predictions and calculations.

The idea of prudence has benefited from the criticism of the tradition of scientific rationalism, which was addressed by David Hume (1896). Hume admitted that reason can discover relations between ideas; that is, it can demonstrate that a certain conclusion follows if a given premise is postulated. However, he assumed that reason alone is unable to demonstrate the *a priori* truth of any matter of fact. Nor can relationships between matters of fact be grounded on causality. Relationships between cause and effect, derived empirically through a certain degree of regularity, cannot be regarded as a rational necessity. In ethics, Hume questioned the role of reason in ethical decision-making. He considered reason as instrumental to human conduct as it can show means for achieving a given end, but it cannot demonstrate whether the end, itself, is desirable or not, as it is subjected to human desire and passion: 'Morals excite passions, and produce or prevent action' (Hume, 1896, Book III, part I, sec. I). Because of that, reason cannot be employed by mathematical or empirical sciences models in politics as politics is inevitably engaged in issues of sentiment and emotion deeply entrenched in habits, customs and predictable ways of satisfying human purposes. Political behaviour should be explained in terms of prudential rather than scientific or technical knowledge.

Edmund Burke has further developed Aristotle's idea of prudence in political ethics. For Burke, political ethics is ruled by prudence rather than by abstractions and universals:

'A statesman differs from a professor in a university. The latter has only the general view of society; the former, the statesman, has a number of circumstances to combine with these general ideas, and to take into consideration. Circumstances are infinite, are in finitely combined, are variable and transient. He who does not take them into consideration is not erroneous, but stark mad [...] he is metaphysically mad. A statesman, never losing sight of principles, is to be guided by circumstances, and judging contrary to the exigencies of the moment, he may serve his country forever' (Burke, 1887, p. 457).

Burke demonstrated his idea of prudence in suggesting a conciliation towards the American colonies. Following Burke, a British statesman should deliberate on the circumstances and character of the American people instead of on dealing with legal or speculative adjustments to maintain the unity of the empire and the legal basis of legislative power: 'I was persuaded that government was a practical thing made for the happiness of mankind and not to gratify the schemes of visionary politicians. Our business was to rule and not to wrangle. And it would have been a poor compensation that we had triumphed in a dispute, whilst we lost an empire' (Burke, 1887, p. 459). Burke's belief in prudential wisdom rejects the grounding of scientific rationalism and utilitarian and pragmatic approaches to politics that are based on a scientific calculus of individual pains.

A utilitarian perspective on rationalism claims to justify a sufficient basis for political action if we understand policy preferences as requiring public justification (Stoker, 1992, p. 372). Accordingly, an individual's rational behaviour is derivative of the considerations of the collective. Political action aimed at advancing one's self-interested preferences is justified best as achieving collective well-being. Therefore, the maximizing political act (if there is just one) is the possible action with the highest expected utility. That general contention can ostensibly be justified because one's self-interested preferences should be merged with others' preferences to achieve the collective best outcome. This utilitarian justification conveys some tension as it requires reconciling self-interests with the interests of others as part of being engaged in community life. This tension is recognized by Brock in his discussion of morality and public justification:

> Morality in the narrow sense: includes all those principles that restrict the individual's personal goals and the advancement of his self-interest. Morality in the broad sense is "the art of life", that is, the precepts instructing people as to how to live, and what makes for a successful, meaningful, worthwhile life.
>
> (Brock, 1998, p. 570)

Confronted with the confining role of public justification in politics, instrumental rationality cannot exhaust political behaviour. It fails to recognize the complexities of social life that may influence the way individuals deliberate about their ends or preferences and the reasons for action and therefore cannot be explained by scientific causation. The imperative concept of prudential rationality is implied by Hans Morganthau: '[p]olitics is an art, not a science, and what is required for its mastery is not the rationality of the engineer, but the wisdom and moral strength of the stateman' (Morgentbau, 1946).

The difficulty with interpreting rationality exclusively in instrumental terms is that it identifies rational action with ethical action from a utilitarian approach. This may ultimately be in the person's self-interest, but it is not necessarily something that increases rational behaviour. The use of instrumental rationality interchangeably with self-interest is evident in Rawls' *A Theory of Justice*. For example, on one page, Rawls offers the interpretation of rationality as self-interest 'I have assumed throughout that the persons in the original position are rational. In choosing between principles each tries as best he can to advance his interests' (Rawls, 1971, p. 142). Yet on the next page, Rawls provides an instrumentalist interpretation to rational behaviour:

> Thus in the usual way, a rational person is thought to have a coherent set of preferences between the options open to him. He ranks these options according to how well they further his purposes; he follows the plan which will satisfy more of his desires rather than less, and which has the greater chance of being successfully executed.
>
> (Rawls, 1971, p. 143)

Sliding between instrumental rationality and self-interest exposes the difficulties encountered by both utilitarian ethics and liberal-rationalist tradition that acting on ethical grounds invariably requires setting aside self-interest. Kurt Baier refers to this difficulty as the 'inconsistent triad' in which rational action can be justified in showing that such action would be for his good; persuasive moral imperatives may force one to act irrespective of whether it would be for his own good; and that moral considerations could not require this if by acting from them meet rational behaviour (Baier, 1986).

Baier suggests that a 'rational' approach to ethics makes it necessary to elucidate the relationships between self-interest, rational choice and ethical principle. It follows that there is no point (again, in a specific sense) in judging self-interest in politics as ethically right or wrong, and that ethical standards of self-interest are applicable to a wide, circumscribable range of political behaviour and motivations as long as it incorporates prudential rather than an instrumental interpretation of rationalism.

Reconciliation project-justification for ethical egoism in politics

Difficulties with self-interest, rationality and ethical justification have led ethical egoists to broaden their view of self-interest in such a way that the self is seen morally as a bearer of rational projects. Once self-interest is seen in terms of rational self-interest, it cannot maintain as narrow, calculating outcomes as when it was interpreted in instrumental terms. Project self-interest encompasses a wider range of possible motivations than coherent preferences, including moral duties, and thus seems more compatible with a prudential grounding of rational behaviour. By viewing a prudential approach to rationality as a theory about political behaviour, it does involve flexibility of the concept of self-interest which seems a healthy corrective to instrumental rationalism of utilitarian and liberal tradition. However, the flexibility of prudential rationality grounded in ethical egoism but voids greater abstraction and inclusiveness, which often results in tautology (Gert, 1967). For that, we must provide some limitations on what counts as a self-interested project. These limitations were discussed in Kavka's Reconciliation Project (1985) but first, we must set out the major ways in which ethical egoism has been conceived as a moral action guide and outlook on politics over the past decade (Kavka, 1985). Our aim is to present a conception of ethical egoism which warrants it being recognized as a basis for prudential rationality.

What is ethical egoism?

Ethical egoism is defined as 'the view that human conduct should be based exclusively on self-interest' (Facione, Scherer & Attig, 1978, p. 45). All definitions of ethical egoism underlie its defining characteristic: people should always act in their own interests. These include, among many others, '[e]ach and every man ought to look out for himself alone' (Emmons, 1969, p. 113), '[e]veryone ought to concern himself with his own welfare alone' (Emmons, 1969, p. 112), '[m]y sole duty is to promote my own interests exclusively' (Hospers, 1961, p. 10) and '[e]veryone ought exclusively to pursue his own interests' (Williams, 1973, p. 250).

By drawing on these conceptualizations, ethical egoism constitutes a distinctive theory as it advocates the exclusive pursuit of self-interest, as a sense that (a) the egoist ought to do those actions of which they benefit to himself alone, or (b) that even if doing actions which will benefit himself along with others, his motivation for acting must be to benefit himself. In both (a) and (b), the egoist is defined as pursuing his own interests 'exclusively', while the object of the exclusion differs between these cases. The sense (b) in comparison to (1) provides a wider range of one's interests to be satisfied than does sense (a). This inclusive sense of (b) makes a more plausible claim for justifying an individual's actions in pursuing her interests irrespective of whether these actions will benefit herself alone or others as well (Regis, 1980 p. 52). Baier describes a version of egoism in the broader sense: 'Egoism recommends that we work out for ourselves in the light of our knowledge, or predilections, preferences, likes and dislikes [...] a life plan whose realization would make our life as rich and worthwhile as possible' (Baier, 1986, p. 220).

Reconciliation project: ethics and self-interest reconciled with prudential reasoning

Extending the scope of what counts as self-interest, including altruistic projects seems more compatible with our experience of the range of human motivations but is also at risk of becoming impractical if anything counts as an egoistic project. This challenge has led George Kavka (1985) to place some restrictions on what counts as a self-interest project as part of his 'reconciliation project', namely the proposed reconciliation between ethics, self-interest and prudential reasoning. While morality and self-interest are the explicit subjects of the project, rationality assumes to serve as a framework and final arbitrator within the project. Kavka's endeavor lies in validating the necessity of reasoning from prudential rationality in trying to reconcile morality and self-interest. Kavka considers Hobbes as one of the earliest, most cited thinkers in connection to the *reconciliation project* on the prudential justification for moral action. Hobbes constitutes a state of nature—a set of hypothetical conditions for human life that precede the existence of the social state. In the state of nature, rational agents pursue their self-interests while no social rules or institutions impose restrictions on their actions. Without the existence of constraints, agents' self-interests inevitably conflict, which can result in a constant state of war among self-interested agents. As such, rational agents seek to escape the state of nature. In exchange for natural freedom, they wish to gain security and peaceful coexistence, allowing the enforcement of rules and institutions that mutually constrain everyone's self-interests. Hobbes' argument on the incentive of every man in the state of nature to accept mutual constraints on their self-interest, so as to attain security and peace underlies a prudential basis for rationality.

Hobbes' prudential approach treats ethics as a set of social rules and policies that mutually constrain men's self-interest, so as to provide them with mutual benefits, such as peace, security and cooperation. Following Kavka, Hobbes adopts a 'rationale' to validate moral principles and rules which define and determine self-interest on external sanctions or societal punishments. Hobbes presents an egoist-normative theory because of the contention that an action that advances one's self-interest is not itself invoked in justification; rather, the justification for self-interested action derives from what is actually expected of us— when we fail to fulfill theses expectations we are criticized or punished. Faced with this assumption, morality cannot exist without a sufficiently punitive society in which rationally self-interested agents comply with moral rules to avoid external sanctions.

Kavka identifies some drawbacks in Hobbesian theory solely relying on external sanctions in attempting to reconcile self-interest, morality and prudential rationality. The absence of internal sanctions allows a prudent justification for self-interest agents to violate morality when it will yield a higher payoff, and when they can gain accurate information about the level of risk they may face. Kavka refers to this as a 'paradox of self-interest: being purely self-interested will not always best serve [their] interests' (Kavka, 1985, p. 306).

That is, internal sanctions might create significant advantages for acting morally, even when morality conflicts with one's direct self-interest. In this case, it would be in one's self-interest to become a moral person, such that one is not disposed to act immorally when doing so would be in one's direct self-interest, so as to gain access to these exclusive payoffs. Kavka rejects narrowing down self-interest, while admitting the need to single out some limitations on what counts as self-interested projects. Kavka lists the following goals worth pursuing under prudential rationality: 'the agent's pleasure, pain, wealth, power, security, liberty, glory, possession of particular objects, fame, health, longevity, status, self-respect, self-development, self-assertion, reputation, honor and affection' (Kavka, 1986, p. 42). This modified normative version of ethical egoism excludes self-destructive, action done for the sake of duty or for the sake of others, from the ways in which one might attempt to justify self-interested action as right.

Kavka observes that Hobbes' approach fails to account for the reconciliation of prudence with any altruistic actions. He calls for a modified version of ethical egoism to overcome the problem of justifying non-self-interested motives as rational to act on them: 'The question of whether moral requirements are consistent with the rational pursuit of the actual ends that people have is both distinct from and more important than the question of whether these requirements cohere with the demands of prudence' (Kavka, 1985, p. 315). Incorporating internal sanctions supports the self-regulation of agents' actions by the sanctions they apply to themselves. They act in a way that brings them a sense of self-satisfaction and a sense of self-worth. However, they refrain from acting in ways that disrupt their moral standards because such behaviour will bring self-condemnation. Self-sanctions thus keep conduct in line with internal standards which makes reconciling morality and self-interest more plausible than suggested in Hobbesian theory. Thus, the value of prudential rationality in moral behaviour depends on circumstances, social environments, individuals' capacities for internal sanctions, on information sets and the actions of others.

In practice, the reconciliation project demands self-regarding and moral capacities to widen the scope of self-interest and thus convey more flexibility to rational self-interest project.

The requirements of prudential rationality in the digital-governance era are:

1. Prudential rationality brings space into the analysis of the application of machine learning, cognitive computing, algorithmic, and data governance. Since space is relational, fluid and a constantly evolving concept, it should be used to contextualize self-regarding interests and other regarding interests, both of which have an equal part in AI and Big Data management decisions including the following: What are the implications of these constructions on how technologies are presented? Who has agency in different constructions of public space and private space? How one can demarcate the boundaries between public and private spaces? At the individual level, enlightened self-interest takes place when people act in ways to further the interests of others, which acts ultimately serve their own interests as well.

2. Prudential rationality urges corporations and business practices to use practical wisdom to discern the interests of those who will be affected by those practices and by not disadvantaging those who are already disadvantaged and not weakening the capabilities of people to lead decent lives. Viewed in this way, prudential rationality ethics contributes to an understanding of what constitutes practical reasoning in a digital-governance era by providing a solid moral grounding for a corporation to surpass mere egoism and become a responsible corporation. To make use of this construct, one should refer to rationality as the defining criterion of responsibility.

Public managers can ask questions such as these about whether or not their options and choices abide by the standards of prudential justice: Does the decision, practice or policy promote conditions within which people can exercise their freedom and capabilities and thereby live a decent life?

Asking and answering such questions, based on the requirements of prudential justice, would greatly assist managers and executives in knowing right from wrong and practicing ethical business conduct. The potential for prudential rationality to contribute to a fairer representation of disadvantaged individuals' involvement in designing and engineering algorithmic systems lies in addressing the inequalities experienced by various social groups and the multiple and intersecting disadvantages underlying the constructions of subject positions. Prudential rationality encourages revealing essential information that often remains hidden in the rational decision-making process. It helps to remedy '[s]ituations where people are believed to have been silenced or excluded from decisions which would directly affect them and which do not acknowledge their knowledge or expertise' (Townsend, 2013, p. 36).

In light of these arguments, although we live in a determinist world in terms of technological development, we are rational agents. We must deliberate about how we act, and it is clear that those deliberations and actions make a difference. The need to anticipate future ethical considerations of AI applications cannot avoid the necessity for a practical reason. Even if a decision taken in good faith, and with the wellbeing of participants in mind, could paradoxically result in more harm than good, mirroring some of the patronizing practices that have traditionally hindered the emergence of the perspectives of disadvantaged individuals in society.

References

Aristotle (1962). Nicomachean Ethics. (M. Ostwald, Trans.). Macmillan. (Book VI, chapter 13, 1144b, lines 31–2).

Barben, D., Fisher, E., Selin, C., & Guston, D. (2008). Anticipatory governance of nanotechnology: foresight, engagement, and integration. In E. Hackett, M. Lynch, & J. Wajcman (Eds.), *The Handbook of Science and Technology Studies.*, 3rd ed. (pp. 979–1000) MIT Press.

Baier, K. (1986). Justification in Ethics. In J. Roland Pennock J. W. Chapman (Eds.), Nomos 28: Justification New York University.

Birren, J. E., & Svensson, C. M. (2005). Wisdom in history. In J. S. Robert & J. Jennifer (Eds.), *A handbook of wisdom: Psychological perspectives* (pp. 3–31). Cambridge University Press.

Blok, V. & Lemmens, P. (2015). The emerging concept of responsible innovation. Three reasons why it is questionable and calls for a radical transformation of the concept of innovation. In J. Van den Hoven, E. J. Koops, H.A. Romijn, T. E. Swierstra & I. Oosterlaken (Eds.) Responsible innovation 2: concept, approaches and applications (pp. 19–36). Springer.

Borup, M., Brown, N., Konrad, K., Van Lente, H. (2006). The sociology of expectations in science and technology. *Technology Analysis & Strategic Management, 18*, 285–298.

Burke, E. (1887). *An Appeal from New to Old Whigs.* Works of Edmund Burke. Bell & Sons.

Brown, N., Rappert, B., & Webster, A. (Eds.) (2000). *Contested Futures: A Sociology of Prospective TechnoScience.* Ashgate.

Brown, N., & Michael, M. (2003). A sociology of expectations: retrospecting prospects and prospecting retrospects. *Technology Analysis and Strategic Management, 15*, 3–18.

Brock, D. W. (1998). *Paternalism and Autonomy. Ethics, 98*, 570.

Callon, M., Lascoumes, P., Barthe, Y., (2009). *Acting in an Uncertain World: An Essay on Technical Democracy.* MIT Press.

Chilvers, J. (2013). Reflexive engagement? Actors, learning, and reflexivity in public dialogue on science and technology. *Science Communication, 35*(2), 283–310.

Douglas, H. (2003). The moral responsibilities of scientists (tensions between autonomy and responsibility). *American Philosophical Quarterly, 40*, 59–68.

Dryzek, J. (2011). *Foundations and Frontiers of Deliberative Governance.* Oxford University Press.

Emmons, D. (1969). Refuting the Egoist. *Personalist, 50*, 309–19.

Facione P. A., Donald S., & Thomas A. (1978). *Values and Society.* Prentice-Hall, Inc.

Felt, U., & Wynne, B (2007). Taking European Knowledge Seriously. *Report of the Expert Group on Science and Governance to the Science, Economy and Society Directorate.* Directorate-General for Research. European Commission, Retrieved December 12, 2020, from: http://ec.europa.eu/research/science-society/documentlibrary/pdf06/european-knowledge-societyen.pdf

Fortun, M. (2001). Mediated speculations in the genomics futures markets. *New Genetics and Society, 20*, 139–156.

Fortun, M. (2005). For an ethics of promising, or: a few kind words about James Watson. *New Genetics and Society, 24*, 157–174.

Gert, B. (1967). Hobbes and Psychological Egoism. *Journal of the History of Ideas, 28* (4), 503–520.

Goodin, R., & Dryzek, J. (2006). Deliberative impacts: the macro-political uptake of mini-publics. *Politics & Society, 34*, 219–244.

Gomila, A., & Amengual, A. (2009). Moral emotions for autonomous agents. In J. Vallverdu & D. Casacuberta (Eds.), *Handbook of Research on Synthetic Emotions and Sociable Robotics: New Applications in Affective Computing and Artificial Intelligence* (pp. 161–174). IGI Global.

Grin, J., & Grunwald, A. (Eds.) (2000). *Vision Assessment: Shaping Technology in 21st Century Society. Towards a Repertoire for Technology Assessment.* Springer.

Grinbaum, A., & Groves, C. (2013). What is "responsible" about responsible innovation? Understanding the ethical issues. In R. Owen, J. Bessant, & M. Heintz (Eds.), *Responsible Innovation: Managing the Responsible Emergence of Science and Innovation in Society* (pp. 110–142). Wiley.

Guston, D. (2012). The pumpkin or the tiger? Michael Polanyi, Frederick Soddy, and anticipating emerging technologies. *Minerva, 50,* 363–379.

Henwood, K., & Pidgeon, N. (2013). *What is the Relationship between Identity and Technological, Economic, Demographic, Environmental and Political Change Viewed through a Risk Lens?* Government Office for Science, Retrieved December 12, 2020, from: www.bis.gov.uk/assets/foresight/docs/identity/13-519-identity-andchange-through-a%20risk-lens.pdf

Hospers, J. (1961). Baier and Medlin on Ethical Egoism. *Philosophical Studies, 12,* 10–16.

Hume, D. (1896). *A Treatise on Human Nature.* Clarendon Press.

Hunt, S. D., & Vitell S. J. (1993) The General Theory of Marketing Ethics: A Retrospective and Revision. In N. C. Smith & J. A. Quelch (Eds.), *Ethics in Marketing* (pp. 775–784) Irwin.

Irwin, A. (2006). The politics of talk: coming to terms with the 'new' scientific governance. *Social Studies of Science, 36,* 299–330.

Irwin, A., Jensen, T., Jones, K., (2013). The good, the bad and the perfect: criticizing engagement practice. *Social Studies of Science, 43,* 118–135.

Jasanoff, S. (2003). Technologies of humility: citizen participation in governing science. *Minerva, 41,* 223–244.

Kant, I. (1785/1988). *Fundamental Principles of metaphysics of morals* (T. K. Abbot. Trans.). Prometheus Books.

Kavka, G. S. (1985). The Reconciliation Project. In D. Copp & D. Zimmerman (Eds.) *Morality, Reason and Truth: New Essays on the Foundations of Ethics* (pp. 297–317). Rowman and Allanheld.

Kavka, G. (1986). Hobbesian Moral and Political Theory. *Studies in Moral, Political, and Legal Philosophy.* Princeton University Press.

Latour, B. (2008). 'It's development, stupid!' or: How to modernize modernization, In J. Proctor (Ed.), Post-environmentalism. MIT Press.

Lövbrand, E., Pielke, R., & Beck, S. (2011). A democracy paradox in studies of science and technology. *Science, Technology & Human Values, 36,* 474–496.

MacIntyre, A. (1985). *After Virtue: A Study in Moral Theory,* 2nd ed. Duckworth.

Macnaghten, P., & Szerszynski, B. (2013). Living the global social experiment: an analysis of public discourse on solar radiation management and its implications for governance. *Global Environmental Change, 23,* 465–474.

Macnaghten, P., & Chilvers, J. (2014). The future of science governance: Publics, policies, practices. *Environment and Planning, C 32*(3), 530–548.

Martin, B. (2010). The origins of the concept of 'foresight' in science and technology: an insider's perspective. *Technological Forecasting and Social Change, 77,* 1438–1447.

Merrill, J. C. (2011). Theoretical Foundations for Media Ethics. In A. D. Gordon, J. M. Kittross, J. C. C. Merrill, W. Babcock, & M. Dorsher (Eds.) *Controversies in Media Ethics*, 3rd ed. (pp. 3–32).Routledge.

Mill, J. S. (1871a). *Utilitarianism.* in J. M. Robson (Ed.), *Collected Works.* University of Toronto Press (1969).

Miller, C., & Bennett, I. (2008). Thinking longer term about technology: is there value in science fiction-inspired approaches to constructing futures? *Science and Public Policy, 35*, 597–606.

Mitcham, C. (2003). Co-responsibility for research integrity. *Science and Engineering Ethics, 9*, 273–290.

Moore, A. (2010). Beyond participation: opening up political theory in STS. *Social Studies of Science, 40*, 793–799.

Morlacchi, P., & Martin, B. (2009). Emerging challenges for science, technology and innovation policy research: a reflexive overview. *Research Policy, 38*, 571–582.

Morgentbau, H. (1946). *Scientific Man Versus Power Politics*. University of Chicago Press.

Nathan, G. (2015). Innovation process and ethics in technology: An approach to ethical (responsible) innovation governance. *Journal on Chain and Network Science, 15*(2), 119–134.

Osbeck, L. M., & Daniel N. R. (2005). "Philosophical Theories of Wisdom." In R. J. Sternberg & J. Jordan (Eds.), *A Handbook of Wisdom: Psychological Perspectives* (pp. 6–83).Cambridge University Press.

Owen, R., Macnaghten, P., & Stilgoe, J. (2012). Responsible research and innovation: from science in society to science for society, with society. *Science and Public Policy, 39*, 751–760.

Pan, Y., & Sparks, J. R. (2012). Predictors, consequence, and measurement of ethical judgments: Review and meta-analysis. *Journal of Business Research, 65*(1), 84–91.

Pavie, X., Scholten V., & D. Carthy. (2014). *Responsible innovation: from concept to practice*. World Scientific.

Pellizzoni, L. (2004). Responsibility and environmental governance. *Environmental Politics, 13*, 541–565.

Rawls J. (1971). *A Theory of Justice*. Harvard University Press.

Regis Jr. E. (1980). What is ethical egoism? *Ethics, 91*(1), 50–62.

Robinson, D. (2009). Co-evolutionary scenarios: an application to prospecting futures of the responsible development of nanotechnology. *Technological Forecasting and Social Change, 76*, 1222–1239.

Robinson, D. N. (1990). Wisdom through the Ages. In R. J. Sternberg (Ed.), Wisdom: Its Nature, Origins, and Development (pp. 13–24).Cambridge University Press.

Rowe, G., & Frewer, L. (2000). Public participation methods: a framework for evaluation. *Science, Technology & Human Values, 25*, 3–29.

Rowe, G., & Frewer, L. (2005). A typology of public engagement mechanisms. *Science, Technology & Human Values, 30*, 251–290.

Schot, J., & Geels, F. (2008). Strategic niche management and sustainable innovation journeys: theory, findings, research agenda, and policy. *Technology Analysis & Strategic Management, 20*, 537–554.

Schuurbiers, D. (2011). What happens in the lab: applying midstream modulation to enhance critical reflection in the laboratory. *Science and Engineering Ethics, 17*, 769–788.

Schuurbiers, D., & Fisher, E. (2009). Lab-scale intervention. Science and society series on convergence research. *EMBO Reports, 10*, 424–427.

Selin, C., 2007. Expectations and the emergence of nanotechnology. *Science, Technology & Human Values, 32*, 196–220.

Selin, C. (2011). Negotiating plausibility: intervening in the future of nanotechnology. Science and Engineering. *Ethics, 17*, 723–737.

Smith, A., Stirling, A., & Berkhout, F. (2005). The governance of sustainable sociotechnical transitions. *Research Policy, 34*(10), 1491–1510.

Steneck, N. (2006). Fostering integrity in research: definitions, current knowledge, and future directions. Science and Engineering. *Ethics, 12*, 53–74.

Stilgoe, J., Owen, R., & Macnaghten, P. (2013). Developing a framework for responsible innovation. *Research Policy, 42*(9), 1568–1580.

Stirling, A. (2006). Analysis, participation and power: justification and closure in participatory multi-criteria analysis. *Land Use Policy, 23*, 95–107.

Stirling, A. (2008). "Opening up" and "closing down": power, participation, and pluralism in the social appraisal of technology. *Science Technology & Human Values, 33*, 262–294.

Stoker, L. (1992). Interests and Ethics in Politics. *American Political Science Review, 86*(2), 369–379.

Te Kulve, H., & Rip, A. (2011). Constructing productive engagement: Pre-engagement tools for emerging technologies. *Science and engineering ethics, 17*(4), 699–714.

Townsend, A. M. (2013). *Smart Cities: Big Data, Civic Hackers, and the Quest for a New Utopia.* W.W. Norton & Company.

Urmson, J. (1998). *Aristotle's Ethics.* Blackwell.

Von Hippel, E. (2005). *Democratizing Innovation.* Retrieved December 12, 2020, from: http://web.mit.edu/evhippel/www/democ1.htm

Von Schomberg, R. (2013). A vision of responsible research and innovation. In R. Owen, J. Bessant, & M. Heintz (Eds.), *Responsible innovation* (pp. 51–74). John Wiley.

Von Schomberg, R. (2011). On identifying plausibility and deliberative public policy. Science and Engineering. *Ethics, 17*, 739–742.

Wilsdon, J., & Willis, R. (2004). *See-Through Science.* Demos.

Williams, B. (1973). *Problem of the Self.* Cambridge University Press.

Wynne, B. (2011). Lab work goes social, and vice-versa: strategising public engagement processes. Science and Engineering. *Ethics, 17*, 791–800.

4 Prudential ethics and intersectionality

Towards a normative framework for AI applications

The aim of this chapter is to bring prudential ethics into conversation with the theoretical laying of intersectionality. Intersectionality provides a compelling framework that can be used to address biases and problems built into existing AI applications, as well as to uncover unique forms of discrimination. Since intersectionality treats social identities as relational, and makes visible the multiple positioning that constitutes everyday life and the power relations that are central to it, it can transform the descriptive and prescriptive accounts of prudential ethics.

As seen in the previous chapter, along with our lengthy review of algorithmic ethics, social identifiers and decision-making become two of the major concerns posed by machine learning algorithms. Social identifiers are pervasive especially when it comes to machine learning algorithms that can learn their correlates when trained on past data. Algorithms are able to discriminate on the basis of a social category, intentionally and unintentionally even if they are not fed social category data, as Korff expresses in his report for the Council of Europe (2013):

> Profiling thus really poses a serious threat of a Kafkaesque world in which powerful corporations and State agencies make decisions that significantly affect their customers and citizens, without those decision makers being able or willing to explain the underlying reasoning for those decisions, and in which those subjects are denied any effective individual or collective remedies. That is how serious the issue of profiling is: it poses a fundamental threat to the most basic principles of the Rule of Law and the relationship between the powerful and the people in a democratic society.
>
> (Korff & Browne, 2013, p. 21)

DOI: 10.4324/9781003106678-5

Intersectionality has spread across disciplinary research areas in recent decades (Holvino, 2010), thus offering criteria and conditions for an effective AI ethical governance framework. The potential for bias and discrimination arising from the use of AI techniques has been addressed by Article 14 ECHR which specifies that rights and freedoms set out in the Convention shall be '[s]ecured without discrimination on any grounds such as sex, race, colour, language, religion, political or other opinions, national or social origin, association with a national minority, property, birth or other status'. Although opportunities for bias are built into the models upon which AI systems are developed (for example, biases inherent in the data sets used to train the models) and when AI systems are implemented in real-world settings, biases can also result from decisions that are systematically biased against groups that have historically been socially disadvantaged (and against individuals who are members of those groups), thereby intensifying unfair discrimination and structural disadvantage (Barocas & Selbst 2016; Wagner Study, 2017, p. 27–28). According to Angwin et al. (2016) these concerns have been raised in association with the use of machine learning techniques to inform custody and sentencing decisions within the US criminal justice system. Allegations criticized the fact that AI techniques operate in ways that are substantially biased against black and other racial minorities.

Therefore, in order to reduce bias in algorithmic decision-making, foregrounding the experiences of particular identity groups and understanding how social identities intersect and which intersectional identities are most sensitive must be addressed in the AI governance framework. For that, we offer to deploy an intersectional approach to ethics thus enabling us to analyse the complexity of subject formations, differences and vehicles of power. This framework of analysis can lead to a form of political praxis based on critique, whereby critique of power often undermines the ways in which interactive processes differently and differentially constitute relations of privilege and penalty.

Thus, the purpose of this chapter is to demonstrate the need for an intersectionality approach to prudential ethics and to discuss how the junction of prudential ethics and intersectionality theory and AI allows us to conceptualize and harness, for the first time, patterns of inequality. To keep with the technological pace while adhering to the public interest, the public sector should acquire the capacities and consider overlapping social categories for supporting and promoting social diversity and inclusion.

As such, this chapter aims to fill a critical void for technological innovation governance scholars and decision makers who have recognized the importance of an intersectionality perspective but who are facing difficulties in understanding its transformational contributions and applications in technological innovation implementation and evaluation processes.

Origins of intersectionality

Development of intersectionality theory

Albeit a fledgling analytical paradigm, the evolution and development of intersectionality has been long in the making. Intersectionality owes its origins to the feminist movement, which sought to develop a more comprehensive, encompassing schema for recognizing and appreciating the converging forces of oppression that affect women on varying levels of identity. Intersectionality arose from the works by women of color in the 1960s as a means for expressing the limitations of feminist theory to accurately portray the struggle for women *across* racial boundaries (Samuels & Ross-Sheriff, 2008, p. 5). In fact, in its most essential form, intersectionality theory is a criticism of second-wave feminism, which was the preeminent mode of feminism in the 1960s in the US. Second-wave feminism expanded the goals set out by first-wave feminism, which primarily took up causes of universalizing suffrage and repealing discriminatory legislation. Second-wave feminism focused on broadening gender equality by addressing issues like sexuality, domestic rights, reproductive rights, and de facto discrimination (Burkett, 2016). However, opponents of the emergent movement asserted that the second wave favored a historical narrative that 'whitewashed' and homogenized the feminist struggle and ignored *voices of difference* from minority communities, such as women of color and queer women (Orr & Braithwaite, 2012).

The term *intersectionality* was first coined by Kimberle Crenshaw (1989) who used the metaphor of intersecting roads to illuminate how differing levels of oppression on the grounds of gender and race interact with one another to create a new, unique experience of marginalization and discrimination. Crenshaw offered the concept of intersectionality as redress to the singularity and one-dimensionality of the consideration of systems of oppression. Although intersectionality theory's origin is rooted in the struggle of women of color for recognition within the big-tent feminist movement of the 1960s and 1970s, as Samuels and Ross-Sheriff (2008, p. 5) note, it went even further and called on scholars to acknowledge that 'for many women of color, their feminist efforts are simultaneously embedded and woven into their efforts against racism, classism, and other threats to their access to equal opportunities and social justice'.

The modern definition of intersectionality holds 'that gender cannot be used as a single analytic frame without also exploring how issues of race, migration status, history, and social class, in particular, come to bear on one's experience as a woman' (Samuels & Ross-Sheriff 2008, p. 5). Consequently, the methodological approaches of researchers and academics employing an intersectional technique mandate that they explore the multitude of 'the overlapping and mutually reinforcing' systems of oppression. The once-accepted universalist approach to the constructs of 'woman' or 'feminist' as singular, all-encompassing experiences has now been replaced by analyses that consider women as whole individuals whose identities may be informed and reinforced by multiple interlocking structures of oppression. Finally, current intellectual pursuits of intersectional analyses incorporate not only mutually reinforcing systems of oppression, but also the myriad of privileges that inform feminist experience. Because intersectionality was developed in reaction to and as a criticism of the tendency of feminist narratives to whitewash oppression experiences, the development of intersectionality theory evolved alongside the evolution of the feminist movement.

Just as important as the *subjects* of intersectional analysis is the relationship between the myriad of systems of oppression and structures of power which interact to shape intersectional experiences. This paradigm states that just like the different identity layers coalesce to create unique experiences of discrimination, the structures of domination that perpetuate systems of inequality are inherently reliant on one another Fellows and Razack (1997, p. 335). suggest that this mechanism functions as mutually-assimilated networks that 'rely on one another' so that '[s]ystems of oppression could not be accomplished without gender and racial hierarchies; imperialism could not function without class exploitation, sexism, heterosexism and so on'.

In addition to the myriad of levels of oppression, intersectionality acknowledges the varying degrees of privilege that also inform the unique experiences of women. These privileges occur naturally from the deficits created by the structures of oppression. An example of the symbiotic relationship between privilege and oppression is evident in Samuels and Ross-Sheriff's research of black or multiracial young children adopted by white parents. Because there was largely socioeconomic, ethnic, and cultural homogeneity of the interviewees' neighborhoods, the children were inevitably a racial minority in their own communities. Here, the interplay between privilege (socioeconomic status) creates the very incubator in which the biracial children experience a *unique* system of oppression. Being a biracial or trans-racial adoptee in a white community meant that their experience of structural oppression was unique to their particular set of privileges and oppressions. Although having two white parents meant that they were transmitted culture which ultimately usually allowed the adoptees to operate in white contexts comfortably, few of them reported dating in high school because their appearance was devalued by the dominant 'Eurocentric images of beauty' (Samuels & Ross-Sheriff, 2008, p. 7). In this example, we see that while being raised in a white community endows certain privileges, it simultaneously and inherently creates situations of alienation. Samuels and Ross-Sheriff's anecdotal research demonstrates that not only is oppression an integral component of intersectionality, but in order to fully appreciate the extent of systems of oppression on individual experiences, academics must also consider networks of privilege.

Although intersectionality theory is intrinsically related to the feminist movement, the methodological contribution of intersectionality reaches far beyond just this field. The intersectional methodology encourages researchers to investigate the multilayered effects of experiences of oppression in their unique and varied manifestations. This is a departure from traditional methodological techniques that often pursue a parsimonious quality in both variables and conclusions. The epistemological approaches to the different forms of oppression to this point had been discrete in nature—the exploration of the patriarchy was a distinct pursuit, and therefore experiences of victimization from institutionalized sexism were interpreted as if they existed in a vacuum. Likewise, racism was investigated as a stand-alone system of persecution. In many ways, the methodological inadequacies of research had a deterministic effect on the analysis of the experiences of oppression themselves. Crenshaw put forth that a 'single-axis framework' failed to consider the compounded marginalization that women of color faced. Crenshaw's foundation offered a theoretical schema for understanding not only bi-level discrimination, but multiple layers of oppression (Dhamoon, 2011, p. 231). This differed dramatically from the traditional single-group approach, which attempted to investigate phenomena '[b]y analyzing the intersection of a subset of dimensions of multiple categories' (McCall 2005, p. 1787). Single-subject design is a subgroup of the categorical comparative approach and is useful for streamlining analytical spaces which can become convoluted when multiple groups and levels are compared side-by-side (McCall, 2005, p. 1786). For example, if researchers want to compare specific ethnic groups within broader racial classifications—Vietnamese, Thai, and Laotiansubgroups within the more general grouping of Southeast

Asians—it becomes necessary to restrict the breadth of analysis for the sake of comprehension. Therefore, a study of this nature would consider these Southeast Asian subgroups independent of gender or class. Naturally, this method has its advantages as it allows researchers to simplify the subject of their research for 'big picture analysis'. However, the very aspect which makes this analytical framework attractive—the ability to disregard intermediary layers of analysis—is also its pitfall. Research which isolates its subject from the multitude of intervening affective torrents of complexity is ultimately reductionist in its analysis.

Indeed, the widely accepted methodological philosophies in most journals laud frugal linear models that shy away from the 'hierarchical' or 'multilevel' modeling that is emblematic of complexity. Yet, how can a study claim to represent authentic paradigms of real-life phenomena if it ignores the systems of complexity which contribute to those phenomena? Intersectional analysis instead refuses to disregard multilevel analysis and welcomes complexity as the natural eventuality of realistic analytical paragons. The previous example ignored gender and class in the investigation of Southeast Asians in order to maintain simplicity. However, an intersectionality study of Southeast Asians would employ an ecological model to identify the integrative nature of the myriad of Southeast Asian experiences. For example, the *middle-class, Vietnamese man* in comparison to the *lower-class, Vietnamese woman*, and so on and so forth. Although this type of multidimensional, 'interaction effect' modeling makes the research exponentially more complicated, it is arguably the only design equipped to deal with the confluence of multiple systems of oppression and paradigms of power. Furthermore, intersectionality allows researchers to investigate and highlight how relationships' 'multiple and differing sets of interactive processes and systems vary at different levels of life and across time and space' (Dhamoon, 2011, p. 237). This ideation of subjects of oppression and power as dynamic, multilayered and complex is, of course, antipodal to the positivist tradition which assumes that all phenomena are fixed, generalizable and fully conceivable. The latter epistemology, which often relies on categorizing and analyzing subjects based on identity, often collapses into the reductive and derivative analysis. Instead, intersectionality values unpacking and evaluating processes and systems (Dhamoon, 2011).

At the forefront of methodological approaches that strive to satisfy this call for complexity is *anticategorical complexity* (McCall, 2005, p. 1773). This approach deconstructs the reductionist analytical categories, maintaining that social life and social structures are infinitely too complex and dynamic to be fettered by fixed categorical definitions. Anticategorical complexity has been particularly integral in deconstructing once-finite categories such as sexuality and gender and examining how they are instead socially-designed constructs (Fotopoulou, 2012, p. 21). The anticategorical approach was birthed from a moment of critique of the tendency of white, big-tent feminists to frame *women* and *gender* as essential and homogenous categories of all women (McCall, 2005, p. 1776). The crux of the criticism was that no solitary category could aptly account for the host of experiences of the individual. Additionally, most intersectional experiences did not fit cleanly into these socially constructed categories. Critics also highlighted that the pro-categorization camp was reinforcing inequalities by excluding experiences that did not fit comfortably into the socially eschewed constructions.

A second approach to complexity is referred to as *intercategorical complexity* (McCall, 2005, p. 1773). This approach accepts the socially constructed categories pro tempore as a provisional means for tracking the disparities between social groups along multiple lines of intersecting identities, dimensions and power structures. The fundamental assumption of intercategorical complexity is that although the relationships and interstices of inequality are fluid and ever-shifting, by adopting categories and simultaneously considering their intersections, researchers are afforded the leverage granted by comparative modes of analysis (Bauerband & Galupo, 2014). McCall puts forth *intracategorical complexity* as the third approach to the complexity of intersectionality (2005, p. 1773). Intracategorical complexity falls somewhere in between the anticategorical approach, which whole-heartedly rejects categorization, and the intercategorical approach, which provisionally excepts categories, if only for the purpose of comparative analysis. The intracategorical complexity approach appreciates the mythological potential of categories but tends to focus on '[n]eglected points of intersection—"people whose identity crosses the boundaries of traditionally constructed groups"' (Dill, 2002, p. 5; McCall, 2005, p. 1774).

The contribution of intersectional methodologies, although they introduced new obstacles, was useful for the poststructuralist movement and the larger popular movement to deconstruct social boundaries as a means of combatting inequality. Ultimately the methodological subgroups challenged the then-predominant mode of analysis which suffered from a blatant failure to reflect loci of neglected experiences of oppression. In the context of feminist history, this paradigm is evident in the diffusion of the experience by black women via their absorption into the homogenizing category of *Women*.

However, the introduction of intersectionality methodology is not without its limitations. Although intersectionality offers a modernized theoretical basis for researching modes of oppression and privilege, it has simultaneously complicated methods of analysis. In fact, the defining aspect of the methodology of intersectionality studies is '[t]he complexity that arises when the subject of analysis expands to include multiple dimensions of social life' (McCall, 2005, p. 1772). Indeed, most scholars have accepted the legitimacy and necessity of intersectionality to convey the intricacies of intersecting experiences of real life, and yet, there lingers a lack of agreement as to *how* to deploy intersectionality. Although it is wholly intuitive that the research methods and processes would mirror the complexity of real life, it poses pragmatic challenges to researchers striving for the simplicity and cleanliness of theory. McCall (2005) highlights the methodological struggles of feminist researchers and writers by pointing out that traditionally they have focused on discrete methodologies (e.g., ethnography, genealogy, deconstruction) that may not lend the breadth necessary for intersectional studies. However, to be completely exhaustive in analysis poses a task insurmountable in the face of the sheer complexity of social relations and dynamics. Although the inherent demand to satisfy the vast complexity of social structures and relations remains an ever-present challenge for experts in the field of intersectionality, it is unquestionably the best tool for exploring the intricacies of modern phenomena.

More recently, Reyes (2017) and Moore (2012) advocate intersectionality as a useful lens for the shifted focus of code-switching to address marginalized factions' interests within society. This stream of research is especially relevant to the social identity framework of intersectionality underlying the conceptualization of code-switching offered in this chapter. We elucidate the developments of big data by reference to the basic premises of social interactionism. They include the ideas that capturing reality as people sense it is a social production; actors constantly affect one another through their interactions over time; human beings are capable of deliberated actions and the way we interact with others and within ourselves, that we define what exists and decide how to act accordingly. Therefore, we consider social identity as '[t]he self as reflexively understood by the individual in terms of his or her biography' (Giddens, 1991, p. 244). It should be noted that while a person's concept of self may remain consistent over time, social identity is more fluid with a process of shifts and adjustments as it plays out in everyday life. Through a process of social interaction, we work to communicate our identities to others and attribute identities to them (Charon, 2010; Gecas, 1982).

Enhancing prudential ethics with intersectionality

To better understand how prudential ethics can provide a normative framework by integrating the insights of intersectionality, it is useful to sketch out a multi-strand working model for applying intersectionality in ethics and policymaking developed by Parken and Young (2007). This model was developed as a road map for promoting equality in the United Kingdom covering issues related to gender, disability, race, sexual orientation, age and religion (in training and education). The aim of this project was to introduce '[a]n approach that can incorporate and manage the differences in origin and outcomes between strands' (Parken & Young, 2007, p. 26), thus '[a]ttending to the specificities of different forms of equality within a single framework poses considerable difficulties' (Ben-Galim, Campbell, & Lewis, 2007, p. 21). According to Parken and Young,

> Our research began from the premise that what was required was an inclusive method capable of promoting equality through policy design, informed by evidence. We have created a multi-strand approach, which avoids 'strand' issues but values the different knowledge and approaches of 'strand' voices.
>
> (Parken & Young, 2007, p. 29)

As a method to integrate intersectionality into the policymaking process, the multi-strand model combines ethics knowledge and policy knowledge. The first step involves identifying a policy field, in this case AI technologies, and exploring issues within that field from the perspective of each strand. A strand may include adverse risks to society that the application of AI may pose.

The second step involved is to map information about each strand within the field of AI technologies. For example, this involved (a) analyzing qualitative and quantitative data from secondary sources (e.g., technical and scientific data, the ethical and legal knowledge) to determine who bears responsibility and accountability for managing and mitigating them; (b) examining current policies and ethical guidelines that may have an impact directly or indirectly on stakeholders; and (c) reviewing risk and harms assessment findings from various stakeholder groups which are often found in a narrowly focused strand.

The third step involves 'visioning', or, as articulated by Parken and Young (2007, p. 10), '[a]sking ourselves what we can do to make transformative change by creating policy or services that will promote equality and human rights'. This stage of gathering data from the mapping process was conducted for each strand to identify commonalities among strands. The rationale behind such comparisons is to identify common solutions that will benefit all strands.

The fourth step involves 'road testing', aimed at reviewing proposed policy solutions to address identified ethical issues raised in each strand offered by key stakeholder groups. This stage requires the arrangement of consultation/engagement with key stakeholders to reflect on the intended consequences of the proposed policy solutions.

The fifth and final step, 'monitoring and evaluation', involves the policy assessment process including the need to identify and measure policy implementation and performance. The multi-strand model thus involves a range of expertise in AI technology, policy, ethics and human rights and is intended to engage with all relevant stakeholders.

Taken together, the multi-strand model of applying intersectionality in both ethical and policy decision-making processes provides a more comprehensive framework that meets prudential ethics requirements including seeking self-interest within the bonds of communities, developing a more nuanced and complex account of power to challenge existing inequalities and examining whether a policy decision or practice promotes conditions with which individuals can exercise their freedom and capabilities and thereby live a decent life. The prudential rationality approach to ethics might deserve more consideration as it confronts the inequalities experienced by various social groups and the multiple and intersecting disadvantages underlying the constructions of subject positions. Prudential rationality requires a commitment to reveal relevant information that often remains hidden in a rational decision-making process so that the decision-maker should be less fount of expert knowledge and more facilitator of '[s]ituations where people are believed to have been silenced or excluded from decisions which would directly affect them and which do not acknowledge their knowledge or expertise' (Townsend, 2013, p. 36).

To illustrate this integrative approach to AI ethics, we may use the case of Amazon's use of a secret AI recruiting tool that showed bias against women. In 2014, Amazon's team in their Edinburgh office created artificial intelligence software which could filter through resumes to select the most-qualified applicants. According to individuals involved in the project, although the computer program was supposed to mechanize the search for the best talent, Amazon's machine-learning specialist realized that the recruiting tool was discriminating against women (Dastin, 2018). Amazon's use of AI and automation have been the catalysts for Amazon's rise to success in the e-commerce sector, and the corporation tried to apply the same model to its hiring mechanism, in an effort to simplify and streamline the process. The program, which rates applicants according to a five-star system, employed 500 computer models to detect around 50,000 keys in order to recommend the most fitting candidates. A source close to the project said that the goal was to create '[a]n engine where I'm going to give you 100 resumes, it will spit out the top five, and we'll hire those' (Hamilton, 2018). However, it was determined in 2015 that the program discriminated against women for software developer jobs and other technical posts.

The investigation into what caused the AI's bias towards men revealed that its algorithm had been built on AI, which, during its learning process, had sifted primarily through resumes that had been submitted to Amazon over a 10-year period. As a result, the algorithm, which was developed to make decisions based on the knowledge it accrued by looking through historical data, perpetuated the existing biases against women from the previous hiring patterns at Amazon. When those involved in the project realized that the AI was discriminating against women, they adjusted it; however, they ultimately scrapped the project for fear that other latent biases may have been unintentionally imprinted on the program (Vincent, 2018). A source at Amazon responding to *The Verge* indicated that they decided not to move forward with the project for several reasons, among them, the gender bias and that the software had never been implemented in a formal capacity for candidate evaluation. Although in this instance, the developers caught the discriminatory software and abandoned the project before it had been assimilated into the official hiring policy, the dangers of bias in AI pose a very real challenge.

Researchers have been vocal about the potentially detrimental effect that biased mathematical models can have on our lives (Osoba & Welser, 2017). As we grow increasingly more dependent on computer automation, AI is becoming exponentially more integrated into our daily experience. However, the threat of crystallizing and perpetuating our own perceptions and surreptitious biases towards others remains. In an article in the *New York Times* (2016), Kate Crawford wrote that problems of AI bias 'may already be exacerbating inequality in the workplace, at home and in our legal and judicial systems' (Crawford, 2016). Prejudice in the judicial system has already had disturbing effects—according to the *MIT Technology Review*[1] (2019), the use of historical data to train risk-assessment AIs has led software to insert the biases of the past into the hardwiring of the new software. In 2016, a report by Toby Walsh, a professor of AI, revealed that the Correctional Offender Management Profiling for Alternative Sanctions program (COMPAS) incorrectly predicted that there is a higher likelihood of black people being repeat offenders (Walsh, 2017). While AI represents a revolutionary moment in technological development and will inevitably affect every sphere of the human experience, it also raises a host of concerns as to unforeseen implications. Recent trends are a rise in populist attitudes (Inglehard & Norris, 2017), an increase in racism and discrimination and the widening of the wealth gap (Darity et al., 2018). As such, there is growing recognition for the importance of socially and ethically responsible technology. Developers, investors and policy-makers must be proactive about their participation in creating responsible technology in order to safeguard against creating new mechanisms and channels that can reinforce sexism, racism and other forms of discrimination.

There have been some cases where the discrimination has appeared to be innocuous, such as the case of Google photo labeling. Google's automatic digital photo album labeling algorithm was categorizing images of black people as gorillas (Barr, 2015). Although Google apologized and wrote it off as unintentional, the potential for much more serious consequences is clear; the current trajectory indicates that machine-learning algorithms will play a pivotal role in the systems that shape how we are advertised to and inform our decision-making processes. As more processes become automated via machine-learning AI, there are more chances that biases will become assimilated into our technology and inform our everyday experiences. Although Amazon caught its problematic AI before it was implemented, there have been other examples of live gender discrimination on digital platforms. In 2016, experts at Carnegie Mellon University found that women were less likely to be shown ads on Google for prestigious jobs (Gibbs, 2015). These incidents represent the fundamental problem with machine-learning AI—it reflects its creators' biases, both tacit and explicit. Therefore, the development of AI programs, especially those which will be highly integrated into our lives, mandates inclusivity and vigilance against creating software that reinforces extant systems of oppression.

One grassroots movement was founded to do just that. The AI initiative, founded at the Harvard Kennedy School in 2015, makes recommendations for the risks associated with the massive datasets and high-performance computing of the new era of AI. The initiative puts a particular emphasis on creating solutions for the potential disruptive socio-political consequences that will affect '[s]ocial contracts and considerations of agency and personal empowerment' (The AI Initiative, n.d).

It is then suggested that, when prudential ethics expands on understandings of the intersecting factors that shape and determine inequality and vulnerabilities, it offers meaningful attention to how diversity and social identities such as race, class, gender, ability, geography and age interact to form complex inequalities and vulnerabilities. Ironically, digital platforms narrow rather than expand the public dialogue. Data is exchanged among like-minded individuals, reinforcing an exclusive set of world views with no necessity to include others. Algorithm development should consider the foreseeable impacts on members of vulnerable and marginal groups. This may require the collection of more data and the undertaking of intersectionality research methods (Dhamoon 2011; McCall, 2005) that account for the simultaneous operation of various dimensions of inequality and discrimination.

Incorporating intersectionality and prudential ethics into AI policymaking and application

To further illustrate the potential of bringing prudential ethics into dialogue with intersectionality for developing an AI ethical framework, we will apply the multi-strand approach. First, we need to explore ethical issues within the AI technologies field from the perspective of each strand to scrutinize broad dimensions of an ethical framework which has considerable potential in securing responsible innovation governance. For that we offer **three** types of strands that follow from the integration between prudential ethics and intersectionality:

> **Human rights:** The human rights strand of AI is introduced in various inquiries and reports developed by civil society organisations, human rights experts and academic scholarship. For example, the Wagner Study issued by the Committee of Experts on internet intermediaries (MSI-NET) raised adverse implications for human rights of algorithmic decision-making systems that affect rights to a fair trial and due process (Art 6), privacy and data protection (Art 5), freedom of expression (Art 10); freedom of association (Art 11), an effective remedy (Art 13), the prohibition on discrimination (Art 14) and the right to free elections (Art 3, Protocol 1). The report suggests that 'the increasing use of automation and algorithmic decision making in all spheres of public and private life is threatening to disrupt the very concept of human rights as protective shields against state interference. The traditional asymmetry of power and information between state structures and human beings is shifting towards an asymmetry of power and information between operators of algorithms (who may be public or private) and those who are acted upon and governed' (Wagner

Study 2017, p. 33). In addition, the European Group on Ethics in Science and New Technologies (2018) addresses the concerns generated by AI technologies to human rights as leading to a systematic treatment of individuals as objects rather than moral agents: 'AI driven optimisation of social processes based on social scoring systems with which some countries experiment, violate the basic idea of equality and freedom in the same way caste systems do, because they construct "different kinds of people" where there are in reality only "different properties" of people. How can the attack on democratic systems and the utilisation of scoring systems, as a basis for dominance by those who have access to these powerful technologies, be prevented?' [...] Human dignity as the foundation of human rights implies that meaningful human intervention and participation must be possible in matters that concern human beings and their environment. Therefore, in contrast to the automation of production, it is not appropriate to manage and decide about humans in the way we manage and decide about objects or data, even if this is technically conceivable. Such an "autonomous" management of human beings would be unwelcome, and it would undermine the deeply entrenched European core values' (European Group on Ethics in Science and New Technologies, 2018, p. 9–10).

Collective societal values: While uncovering bias generated by AI-enabled systems falls within the scope of existing human rights protection, doing so may undermine important collective interests and values. Indeed, AI applications may generate adverse consequences to collective life as tasks previously performed by humans are automated thus leading to dehumanization and less empathy and human interaction. This goes against the right to meaningful human contact as observed by the Parliamentary Assembly: 'Certain types of robots are equipped with artificial intelligence and are programmed to mimic social abilities in order, for example, to establish a conversation with its user. For instance, care robots can use affective computing in order to recognise human emotions and subsequently adjust the robot's behaviour. Potentially, robots can stimulate human relationships [...] Several studies on the effect of Paro, a soft seal robot, in inpatient elderly care, seem to suggest that the mood of elderly people improves and that depression levels decrease; in addition, their mental condition becomes better, advancing the communication between senior citizens and strengthening their social bonds. However, there is a danger that robots could interfere with the right to respect for family life as an (un)intentional consequence of how the robot affects its users. Due to anthromorphism, vulnerable people such as the elderly may consider a social robot for example as their grandchild. If not treated carefully, the care receiver may focus primarily on the care robot, instead of, for example, his or her family members or other human beings' (Council of Europe Parliamentary Assembly 2017, para 37). It is then suggested that excessive reliance on algorithmic decision-making and AI applications may diminish and devalue opportunities for authentic, meaningful human interactions and relationships, trumped by the promises of greater efficiency, precision and consistency of service that AI applications can offer.

Reciprocity: the adverse consequences arising from the application of AI technologies may result in power asymmetry between those who develop or own algorithmic systems and individual users, groups and populations who are affected by their use. Reciprocity-based appeals to fairness are regulated by accepted rules or procedures that participants accept which enable members of society to realize their own good in ways they regard as fair. As Rawls points out (2001, p. 6): 'Fair terms of cooperation specify an idea of reciprocity, or mutuality: all who do their part as the recognized rules require are to benefit as specified by a public and agreed-upon standard'.

The idea of reciprocity does not necessarily mean that individuals benefit equally; whether equality is required depends on publicly accepted standards. A well-ordered society '[i]s a fair system of social cooperation over time from one generation to the next' (Rawls, 2001, p. 5), in which the '[r]ole of the principles of justice ... is to specify the fair terms of social cooperation' (Rawls, 2001, p. 7). Reciprocity is realized under the recognition of socially-shared priorities embedded in the commitment to live cooperatively with others which demands that one will be '[r]eady to propose principles and standards as fair terms of cooperation and to abide by them willingly, given the assurance that others will likewise do so' (Rawls, 1996, p. 49). John Rawls, in his paramount work, *Justice as Fairness* (2001) proposes that by embracing reciprocity, individuals prompt their commitment to justify one's actions to others on the ground upon which they could not be reasonably discarded. Moreover, Rawls asserts that individuals will not comply with the principles of justice without a reasonable assurance that others will comply as well.

This requirement is magnified by the individualism and import of personal freedom. In this way, Rawls' notion of reciprocity mitigates the American bias towards individualism—encouraging citizens to consider the collective good under the condition that political power should have a justification that meets our shared reason (Thunder, 2006). We, therefore, offer to expand the existing framing of AI risk discourse which is based on the human-rights approach to encompass a prudential approach for addressing collective values and be more responsive to power asymmetry.

Note

1 www.technologyreview.com/magazine/2019/01/

References

Angwin, J., Larson, J., Mattu, S., & Kirchner, L. (2016). Machine bias. *ProPublica*, 23 May. Available at www.propublica.org/article/machine-bias-risk-assessments-in-criminal-sentencing
Barocas, S., & Selbst, A.D. (2016). Big data's disparate impact. *California Law Review 104*, 671.
Barr, A. (2015). Google Mistakenly Tags Black People as 'Gorillas,' Showing Limits of Algorithms. *The Wall Street Journal*. July 1, 2015. Available at www.wsj.com/articles/BL-DGB-42522
Bauerband, L. A., & Galupo, M. P. (2014). The gender identity reflection and rumination scale: Development and psychometric evaluation. *Journal of Counseling & Development*, *92*(2), 219–231.
Ben-Galim, D., Campbell, M. & Lewis, J. (2007). Equality and diversity: A new approach to gender equality policy in the UK. *International Journal of Law in Context 3*, 19–33.
Burkett, E. (2016). 'Women's movement'. In *Encyclopædia Britannica*. Retrieved May 6, 2020, from: www.britannica.com/topic/womens-movement
Charon, J. M. (2010). *Symbolic Interactionism: An Introduction, an Interpretation, an Integration*. Prentice Hall.
Council of Europe, Parliamentary Assembly. (2017). *Committee on Culture, Science, Education and Media. Technological Convergence, Artificial Intelligence and Human Rights*. April 10. Doc 14288. Available at http://semanticpace.net/tools/pdf.aspx?doc=aHR0cDovL2Fzc2VtYmx5LmNvZS5pbnQvbncveG1sL1hSZWYvWDJILURXLWV4dHIuYXNwP2ZpbGVpZD0yMzUzMSZsYW5nPUVO&xsl=aHR0cDovL3NlbWFudGljcGFjZS5uZXQvWHNsdC9QZGYvWFJlZi1XRJlZi1XRC1BVEtY1YTUwyUERGLnhzbA==&xsltparams=ZmlsZWlkPTIzNTMx
Crawford, K. (2016). Artificial intelligence's white guy problem. *The New York Times, 25*.

Crenshaw, K. (1989). Demarginalizing the intersection of race and sex: A black feminist critique of antidiscrimination doctrine, feminist theory, and anti-racist politics. *University of Chicago Legal Forum*, 139–167.

Darity Jr, W., Hamilton, D., Paul, M., Aja, A., Price, A., Moore, A., & Chiopris, C. (2018). *What We Get Wrong About Closing the Racial Wealth Gap*. Samuel DuBois Cook Center on Social Equity. December 12, 2020, from: www.compassworkingcapital.org/dei-blog/test-2/6/7/2018.

Dastin, J. (2018). *Amazon scraps secret AI recruiting tool that showed bias against women*. Reuters.

Dhamoon, R. K. (2011). Considerations on mainstreaming intersectionality. *Political Research Quarterly*, *64*(1), 230–243.

Dill, B. T. (2002). Work at the intersections of race, gender, ethnicity, and other dimensions of difference in higher education. *Connections: Newsletter of the consortium on race, gender, and ethnicity*, 5–7. Retrieved May 6, 2020, from: www.crge.umd.edu/publications/news.pdf.

European Group on Ethics in Science and New Technologies (EGE). (2018). *Statement on Artificial Intelligence, Robotics and 'Autonomous' Systems*. Available at https://ec.europa.eu/research/ege/pdf/ege_ai_statement_2018.pdf

Fellows, M. L., & Razack, S. (1997). 'The race to innocence: Confronting hierarchical relations among women'. *Journal of Gender, Race and Justice, 1*, 335–52.

Fotopoulou, A. (2012). 'Intersectionality queer studies and hybridity: Methodological frameworks for social research'. *Journal of International Women's Studies, 13*(2), 19–32.

Gecas, V. (1982). 'The self-concept'. *Annual Review of Sociology, 8*(1), 1–33.

Gibbs, S. (2015). Women less likely to be shown ads for high-paid jobs on Google, study shows. *The Guardian*. July 1, 2015. Available at www.theguardian.com/technology/2015/jul/08/women-less-likely-ads-high-paid-jobs-google-study

Giddens, A. (1991). *Modernity and self-identity: Self and society in the late modern age*. Stanford University Press.

Hamilton, I. A. (2018). Amazon built an AI tool to hire people but had to shut it down because it was discriminating against women. *Business Insider*. October 10, 2018. Available at www.businessinsider.com/amazon-built-ai-to-hire-people-discriminated-against-women-2018-10

Holvino, E. (2010). Intersections: The simultaneity of race, gender and class in organization studies. *Gender Work and Organization*, *17*, 248–277.

Inglehart, R., & Norris, P. (2017). Trump and the Populist Authoritarian Parties: The Silent Revolution in Reverse. *Perspectives on Politics*, *15*(2), 443–454.

Korff, D., & Browne, I. (2013). 'The use of the Internet & related services, private life & data protection: trends, technologies, threats and implications', *Council of Europe*, T-PD(2013)07. Available at: https://ssrn.com/abstract=2356797

McCall, L. (2005). 'The complexity of intersectionality'. *Signs: Journal of Women in Culture and Society, 30*(3), 1771–800.

McCarthy, J. (1960). *Programs with common sense* (pp. 300–307). Research Laboratory of Electronics (RLE) and MIT computation center.

Moore, M. R. (2012). 'Intersectionality and the study of black, sexual minority women'. *Gender & Society, 26*(1), 33–39.

Orr, C. M., & Braithwaite, A. (Eds.). (2012). *Rethinking women's and gender studies*. Routledge.

Osoba, O. A., & Welser IV, W. (2017). *An intelligence in our image: The risks of bias and errors in artificial intelligence*. Rand Corporation.

Parken, A., & H. Young. (2007). *Integrating the promotion of equality and human rights for all. Cardiff, Wales: Towards the Commission of Equality and Human Rights*. Unpublished report for the Welsh Assembly Government and Equality and Human Rights Commission.

Rawls, J., (1996). *Political liberalism*, Columbia University Press.

Rawls, J. (2001). *Justice as Fairness: A Restatement*. Harvard University Press.

Reyes, P. (2017). 'Working life inequalities: do we need intersectionality?', *Society, Health & Vulnerability, 8* (1), 14–18.

Samuels, G. M., & Ross-Sheriff, F. (2008). 'Identity, oppression and power: Feminisms and intersectionality theory', *Affilia, 23*(5), 5–9.

The AI Initiative. (n.d). December 12, 2020, from: http://ai-initiative.org/

Thunder, D. (2006). A Rawlsian Argument against the Duty of Civility. *American Journal of Political Science, 50*(3), 676–90.

Townsend, A. M. (2013). *Smart cities: Big Data, civic hackers, and the quest for a New Utopia*. W.W. Norton & Company.

Vincent, J. (2018). Amazon reportedly scraps internal AI recruiting tool that was biased against women. *The Verge*. October 10, 2018. Available at Amazon reportedly scraps internal AI recruiting tool that was biased against women.

Wagner, B. (2017). '*Study On The Human Rights Dimensions of Automated Data Processing Techniques (In Particular Algorithms) And Possible Regulatory Implications*'. October 6, 2017. Council of Europe, Committee of Experts on internet intermediaries (MSI-NET). Available at https://rm.coe.int/study-hr-dimension-of-automated-data-processing-incl-algorithms/168075b94a

Walsh, T. (2017). The AI Revolution. *NSW Department of Education: Future Frontiers*. pp. 1–15. Available at https://prod65.education.nsw.gov.au/content/dam/main-education/teaching-and-learning/education-for-a-changing-world/media/documents/The_AI_Revolution_TobyWalsh.pdf

5 Contemporary analysis of AI affairs in the international arena

This chapter will examine the increasing scholastic discourse regarding the non-technological impact of AI on society, including the ethical, social and legal implications on current affairs in the international community. It carries out a critical analysis of current AI and big data affairs in the international arena including Facebook's and Twitter's Fake News Scandals and their aligned government responses, Amazon's use of a secret AI recruiting tool that showed bias against women, mass surveillance systems enacted by governments like the Chinese Social Credit and AI-based monitoring systems during the Coronavirus Pandemic and more. Examining these technological affairs provides the opportunity to re-examine the professional role of the public service and related claims of expertise, public interest, accountability and norms in the cognitive era.

Notably, the evolution of AI and its varied stages raises prevalent ethical and social issues, such as the threat of data bias, automated mass surveillance and possible abuse of computational propaganda. These issues are more likely to aggravate when AI advances and therefore it is of paramount importance to discuss the ethical and legal issues at this point and find suitable coping mechanisms and foster relevant regulatory mechanisms and governance of the design and deployment of AI applications. Nonetheless, there is an increasing discourse regarding the notion of 'AI for Good', as AI has the potential to foster positive developments in an abundant number of human fields. Consequently, the regulators should take the positive and negative impact into consideration and avoid fully preventive policies (De Spiegeleire Maas & Sweijs, 2017).

Broadly, the deployment of AI in varied human spheres poses substantial ethical obstacles with an emphasis on fairness, accountability and transparency. Several scholars argue that there is an urgent need to understand the AI origins and use of training data, models and test

DOI: 10.4324/9781003106678-6

data, as well as the wide environmental and labor costs of AI-driven systems. Notably, the accountability gap is constantly growing and the government regulation regarding AI systems is insufficient to deal with this gap, possibly leading to greater rates of bias and discrimination (Whittaker, Crawford, Dobbe, et al. , 2018).

However, besides the inherent technological loopholes of AI models, (Pasquale, 2015) some scholars argue that there is a salient obstacle to accountability due to the core culture of industrial and legal secrecy in the AI sphere (Ram, 2017; Citron & Pasquale, 2014). Corpocracy secrecy laws make it almost impossible to properly review AI systems, with emphasis on: '[t]raining data, to data models, to the code dictating algorithmic functions, to implementation guidelines and software, to the business decisions that directed design and development' (Whittaker, Crawford, Dobbe, et al. , 2018, p. 11).

Furthermore, AI automated classification is more likely to yield with discriminatory patterns and outcomes. For instance, facial recognition systems do not always detect dark skin people and in several instances they were classified as gorillas (Dougherty, 2015). In addition, one may mention the notable discriminatory affair of Google's search algorithm, which brought biased results, exhibiting occupational gender discrimination (Kay, Matuszek & Munson, 2015).

It is interesting to mention that the policy evolution of AI ethics is derived from the substantial issues mentioned and in recent years receives greater scrutiny from government and non-government actors. Schiff, Biddle, Borenstein and Laas, (2020) discussed the notable AI ethics documents, including policy strategies, norms and frameworks, produced by governments and NGOs and mentioned that since 2016 more than 80 documents were created. Notably, government actors have produced the highest number of documents, including European countries, Canada, China and the United States. There is a prevalent dominance of developed countries in the creation of these documents. This notion raises substantial concerns regarding the involvement of low- and middle- income nations in the establishment of AI frameworks and policies: '[t]here is a risk that AI-driven growth defined and dominated by wealthy countries could detrimentally impact poorer ones'.

In addition to the public sector contribution, the private sector has also produced influential AI ethics documents such as those from Microsoft, Google and IBM. These documents depict the establishment of ethical standards for development, governance strategies and the importance of self-regulation or collective efforts to foster regulation, compared to the traditional government regulation (Microsoft, 2018).

Cultural perspectives of surveillance and facial recognition technologies

In this regard, Hagerty and Rubinov (2019) highlight the importance of regional and cultural differences on AI impact and posit that AI understanding is designed by local cultural and social settings. There is a notable example which occurred in Brazil, as the Portuguese translation of Facebook's like button to 'enjoy' fostered the filtering algorithm to harm indigenous land rights activism in Brazil (Ochigame & Holston, 2016).

In this regard, AI aggravates widespread surveillance in the US, China and other state actors and fosters its alarming automation, as seen in social media tracking, facial recognition, affect recognition and sensor networks. For instance, as further articulated in the AI report:

> The use of affect recognition [...] attempts to read inner emotions by a close analysis of the face and is connected to spurious claims about people's mood, mental health, level of engagement, and guilt or innocence. This technology is already being used for discriminatory and unethical purposes, often without people's knowledge. Facial recognition technology poses its own dangers, reinforcing skewed and potentially discriminatory practices, from criminal justice to education to employment, and presents risks to human rights and civil liberties in multiple countries.
>
> (Whittaker, Crawford, Dobbe, et al., 2018, p. 8)

Notably, AI-driven mass surveillance offers outreaching tracking capacities which exceed the human ability to analyse massive amounts of data and foster predictive capabilities. It is interesting to mention that, AI-driven mechanisms show the inter-state differences in their approach to AI deployment. While some populations may accept AI-driven mass-scale government surveillance, the civil resistance to this phenomenon is salient globally. For instance, in 2016, activists hacked the national ID system in Venezuela, which is based on AI-driven tracking of users in different services, and deleted accounts of politicians as a protest act (Berwick, 2018).

Following the above, one of the most prevalent AI-based developments in the surveillance sphere is the Social Credit System in China. The term refers to a myriad of surveillance initiatives aimed at bolstering the government enforcement of Chinese legislation (State Council, 2020). It is highly publicized and offers mass citizen rankings based on their social conduct and political agenda, building on facial recognition data, extracted from surveillance cameras and social media platforms. Among its goals, one may mention the establishment of a reliable financial credit system, the abolition of market fraud and the decrease of varied 'dishonest' public fields. Interestingly, the surveillance system is well accepted by the Chinese citizens who perceive this mechanism as an efficient vector to tackle fraud. One of the surveys showed that the stronger socio-economic groups are more likely to favor this system, probably as it benefits them (Kostka, 2019).

The Chinese surveillance vision is aimed at monitoring the vast population, with emphasis on marginalized groups. Notable examples include the deployment of facial recognition tools at the Hong Kong-Shenzhen border (Shen, 2018), and the vast use of the Social Credit Monitoring System (Morris, 2018). These tools enable the Chinese government to facilitate far-reaching social control using emerging technologies (Feng & Lucas, 2018; Vanderklippe, 2018). In this regard, there is increasing surveillance in the Xinjiang Autonomous Region, including:

> [p]hysical checkpoints and programs where Uighur households are required to 'adopt' Han Chinese officials into their family, to the widespread use of surveillance cameras, spyware, Wi-Fi sniffers, and biometric data collection, sometimes by stealth. Machine learning tools integrate these streams of data to generate extensive lists of suspects for detention in re-education camps, built by the government to discipline the group.
>
> (Whittaker, Crawford, Dobbe, et al., 2018, p. 13)

Nonetheless, besides facilitating surveillance, the scholarly discourse highlights additional monumental ethical and societal implications of AI. Hagerty and Rubinov (2019) also posit that AI entrenches social cleavages and exacerbates social inequality, especially among marginalized groups. In this regard, one may suggest that developing countries are more likely to be susceptible to the social implications of AI (Barocas & Selbst, 2016). This notion was reinforced by the World Economic Forum Global Future Council on Human Rights, that posited these countries are more likely to suffer from the discriminatory patterns of machine learning (World Economic Forum, 2018). Notably, the aggravation of inequality in geographic regions, where corruption and poverty are incredibly high is an alarming situation.

In addition to the implications on developing economies, one may highlight the prevalent social issues in the US arising from AI evolution, such as job requiring tools that are biased against women (Dastin, 2018), or discriminatory credit algorithms against Latin- and

African-Americans (Bartlett, Adair, Stanton, & Wallace, 2018; Glantz & Martinez, 2018; Waddell, 2016). All of the above demonstrates that AI systems have the potential to diminish the opportunities of historically disadvantaged, marginalized and vulnerable groups.

Following the examples above, there are additional alarming affairs, exhibiting the unprecedented implications of AI systems, including the accusations of triggering genocidal oppression of Myanmar's Rohingya population (Nebehay, 2018; Stecklow, 2018). China's intensive surveillance of local communities (Samuel, 2018; Human Rights Watch; Freedom House, 2018) and the surveillance policies targeting political activists and dissidents by authoritarian regimes (Tazhiyeva, 2018; Kirkpatrick, 2018).

Despite the totalitarian and intrusive reputation of China in the surveillance sphere, one may suggest that democratic actors are also susceptible to employ AI for similar purposes. In this regard, the technological use of surveillance tools by the US government seems to be quite vast, but not as exposed to the public. For instance, the Pentagon promoted research on AI-driven social media tracking to predict mass-scale civil conduct (Nafeez, 2018). Another alarming AI usage was seen with the family separation policy of the US Immigration and Customs Enforcement (ICE) agency. This system combines several data sources to create immigrants' profiles in order to further profile and track individuals (Hao, 2018). According to the *AI Now Report 2018*, the system:

> [f]orcibly removed immigrant children from their parents, employees from Amazon, Salesforce, and Microsoft all asked their companies to end contracts with U.S. Immigration and Customs Enforcement (ICE). Less than a month later, it was revealed that ICE modified its own risk assessment algorithm so that it could only produce one result: the system recommended 'detain' for 100% of immigrants in custody.
>
> (Whittaker, Crawford, Dobbe, et al. , 2018, p.10;
> for further detail see Kastrenakes, 2018;
> Lecher, 2018; Shaban, 2018; Sonnad, 2018)

Among the myriad of AI-driven surveillance tools and methods, racial recognition is considered highly controversial in terms of civil liberties. One may emphasize that facial characteristics are an inherent form of biometric identification, and therefore less likely to change. In this regard, if the facial traits are identified, an individual can be linked to various public and private records, such as criminal records and credit score. Due to the sensitivity of such data, there is an urgent need to moderate any possible tracking of facial recognition data by governments and malign agents. In addition, studies show that facial recognition fosters racial discrimination and bias in the criminal justice system and biased hiring algorithms, as seen in COMPAS's risk assessments and Amazon recruiting tool (Weiner, 2018; Dastin, 2018; Amazon Web Services; 2017; Angwin, Larson, & Kirchner, 2016). Currently, there are varied initiatives aimed at promoting diversity and inclusion, such as Black in AI, Women in Machine Learning, Latinx in AI, and Queer in AI (Whittaker, Crawford, Dobbe, et al. , 2018).

Influence operations and AI threats to the democratic ethos

In addition to the ethical and social implications mentioned above, it is of paramount importance to mention the potential role of AI as a prevalent trigger to the emergence of influence operations and disinformation campaigns. Manheim and Kaplan (2018) elaborate upon the alleged threats of AI to core democratic and human values such as privacy, equality, autonomy and the electoral process. These threats may be reflected within the AI ability to manipulate voter preferences and generate behavioural profiles of individuals. Since social media platforms use AI-driven algorithms to foster their economic growth and revenues, the wide use among users raises substantial threats. In this regard, one of the most prevalent implications of AI progress is the diminishing reputation of democratic institutions (Cohen, 2017).

Some may argue that the core component of AI black-box decision-making (Knight, 2017), including algorithmic bias and its inability to be explainable or fully transparent among users, highly contradicts the democratic and liberal ethos of putting the individual at the center, fully accountable and aware of his or her actions. As much as most scholars do not foresee the dystopian scenario of Singularity and AI replacing humans soon, the current risks to democratic institutions are substantial and require additional regulatory scrutiny (Brundage, 2015). As further articulated by Manheim and Kaplan (2018, p. 1):

> Privacy, anonymity and autonomy are the main casualties of AI's ability to manipulate choices in economic and political decisions. The way forward requires greater attention to these risks at the national level, and attendant regulation. In its absence, technology giants, all of whom are heavily investing in and profiting from AI, will dominate not only the public discourse, but also the future of our core values and democratic institutions.

Broadly, the vast spread of online disinformation has become a constant threat in human society, and advances in AI foster an alarming capacity by state and non-state actors to artificially generate content, mimicking the human narrative and style. Interestingly, the scholars Kreps, McCain and Brundage (2020) examined the credibility of AI-generated content in three different experiments, comparing between original and fake stories. They found that the readers were not likely to differentiate between AI and human-generated content, and that exposure to the content does not impact significantly their policy views. Consequently, this study has severe implications for the vectors malevolent actors might abuse AI functionalities to foster online disinformation and electoral interference, as in Cambridge Analytica. As further articulated in their article:

> [w]e found that people can be manipulated by AI-generated text such that they **cannot discern real from synthetic content**. [...] Malicious actors can easily produce AI-generated content and generate confusion about the truth, **undermining trust in democratic institutions** such as the media. [...] More generally, however, the ease of manipulation suggests avenues for misinformation not in service of political persuasion but instead in **sowing confusion and distrust** [...] the potential to **undermine the basis of coherent public policy**. [...] AI-generated misinformation therefore represents a **plausible future mechanism for the spread of false and misleading information**.
>
> (ibid., p. 4, 6)

Arsenault (2020, p. 4) strengthens these arguments and highlights the substantial challenges of liberal democracies due to disinformation's efforts in the age of AI:

> [t]he distinct threat posed by AI lies in its ability to amass and rapidly analyse huge swaths of data, thereby **providing adversaries with constant analysis and surveillance of public sentiment** [...] the integration of AI into botnet communication has allowed for the automation of disinformation efforts that flood social media with coordinated messages, thereby misrepresenting public debate and **undermining the voter's ability to make political decisions** informed by public sentiment [...] AI's ability to learn from data allows it not only to improve its methods, but to adapt to changing environments, **facilitating the creation of believable propaganda that is almost indistinguishable from human script**.

Nonetheless, even if Russia is considered to be a leading actor employing automated social media bots to augment the pro-Russian agenda and chaos in the Western world, there is no feasible evidence indicating that AI-generated contents have been used to automatically foster electoral interference (Helmus et al., 2018). In addition, the AI function in the disinformation sphere is still highly vague, and not entirely negative. AI also enables to automatically generate, detect and mitigate false content online and impact public opinion. Notably, by offering an innovative and scalable solution to fake news mitigation, such as sophisticated and automated fact checking tools and removal of fake content (Kertysova, 2018).

However, there are notable limitations to the deployment of AI-driven detectors of online disinformation. One may mention the potential threat of over-blocking and even censorship of lawful and legitimate content, as AI is still susceptible to false positives. In this regard, reliable content may be censored due to incorrect AI-labelling, and therefore hamper freedom of expression. Interesting to mention is that AI currently cannot fully evaluate complex semantic structures, which are based upon linguistic nuances and country-driven cultural and political settings. For instance, Ketrysova (2018, p. 60) further explains the semantic loopholes:

> [a]utomated technologies remain limited in their ability to assess the accuracy of individual statements [...] Current AI systems can only identify simple declarative statements, and miss implied claims or claims embedded in complex sentences, which humans recognise easily [...] The same goes for expressions where contextual or cultural cues are necessary.

Following the above, there are several ways in which AI threatens electoral processes and democratic institutions. For instance, weaponized AI can disrupt democratic elections using sophisticate cyber-attacks, weaponized 'micro-targeted' propaganda, or psychological vectors, leading the public to distrust the electoral process. Notably, AI can exacerbate human biases and violates some of the core components of democracies, inducing fairness, accountability and transparency. As recently seen in the international community, AI can foster new opportunities for foreign election interference by state and non-state actors. As further articulated by Manheim and Kaplan (2018, p. 134): '[a]rtificial intelligence can be used both to increase effectiveness and to mask the purposes and methods of voter suppression and manipulation'.

Arsenault (2020) reinforces that notion of AI efficiency and scalability and posits that:

> The proliferation of AI introduces three critical transformations that exacerbate the scope, scale, and efficiency of contemporary disinformation campaigns: AI, which uses advanced algorithms and social media data to precisely target segments of the electorate, provides adversarial actors with a tool for the microtargeted exploitation of pre-existing political fissures and biases [...] AI has allowed for the automation of political propaganda, as exemplified by the use of botnets leading up to elections [...] AI's ability to integrate machine learning and neural network capabilities allows for the production of convincing AI-produced propaganda that seems authentic.

Historically, election interference is not considered novel: 'While false reporting, misdirection, and propaganda are centuries old tactics, artificial intelligence compounds the problem of fake news by making it seem more realistic or relevant through targeted tailoring' (ibid., pp. 144–145). Notably, AI has clearly aggravated the scope and scale of influence operations, as witnessed in the previous electoral scandal of the alleged Russian interference in the 2016 US elections (Macdonald, 2017; Patterson, 2018; Mueller Report, 2019). The Russian influence operation aimed at undermining the public trust in the American government and foster chaos and included 3,841 accounts activated by the Russian Internet Research Agency that spread fake news to reshape the political narrative (Timberg & Harris, 2018). According to the World Economic Forum, AI has already 'silently [taken] over democracy' through the use of social media manipulation, fake news, bots and trolls (Polonski, 2017; Meserole & Polyakova, 2019).

Within the framework of the 2016 US elections, it is essential to elaborate upon the Cambridge Analytica affair. The data science firm created psychological profiles of 230 million Americans and used social media to perform psychological political warfare (Cadwalladr, 2018). Using AI, the firm was able to promote Trump's candidacy and manipulate the voters (Graham, 2020; Miller, 2018) .In this regard, several scholars argue that this affair is monumental, but not time-limited as it is: 'not a one-time event limited to the 2016 election. It's a daily drumbeat. These [fake accounts] are entities trying to disrupt our democratic process by pushing various forms of disinformation into the system' (Roose, 2018).

Manheim and Kaplan (2018) reinforce this notion and suggest the AI spread in additional fields: 'manipulation are hardly the only threats that AI poses to democracy. As more and more public functions are privatized, the scope of constitutional rights diminishes. Further relegating these functions to artificial intelligence allows for hidden decision-making, immune from public scrutiny and control. For instance, predictive policing and AI sentencing in criminal cases can reinforce discriminatory societal practices, but in a way that pretends to be objective. Similar algorithmic biases appear in other areas including credit, employment, and insurance determinations'.

(p. 109)

AI-driven disinformation 2.0 – Deepfake

Following the above, one may mention the AI-driven amplification of disinformation, as reflected by the threat of 'deepfake', which is usually defined as fabricated audio or video aimed at manipulating human senses (Chesney & Citron, 2019). Kertysova (2018, p. 66) adds that it is '[h]ighly realistic and virtually indistinguishable from real material'. Different than the classic methods of machine learning, in which the training set of data samples establish inputs and structures for the algorithm to base on, deepfake posits an amplified threat to human logic and autonomy as it shows the way one can produce video, text and audio forgeries, that contain the online manipulation of speech and visuals (Greg & Chan, 2017).

Digitally, deepfake creation is usually divided into two phases. First, it is trained on large datasets to create neural networks, which are capable of replicating human patterns and rules to mimic similar sorts of media. Following that, the programmers use an additional algorithm called 'discriminator', which reviews the fake content thoroughly in order to '[t]est the deepfake content in order to spot mistakes and improve believability' (Arsenault, 2020, p. 49). Once the fake content is successfully identified, the AI's neural networks can hone their mimicking ability (Chesney & Citron, 2019).

Besides the technological progress, it seems the deepfake threat is amplified for political discourse and civil trust through its ability to seem highly credible. Notably, the literature has highly addressed the role of imagery in the security sphere, where images are more likely to be perceived as neutral and visual facts (Friis, 2015). It is interesting to mention that deepfakes may be designed to trigger an escalated response from a susceptible audience:

[t]here is a concern that deepfakes could be shared in precise, targeted ways; that is, deepfakes will be curated and designed to resonate with those audiences that are most likely to perceive them as 'true' precisely because they seem to 'confirm' or legitimize preexisting suspicions, fears, and ideological commitments ... AI is capable of shaping perceptions, and generating chaos and confusion through the rapid and effective mimicry of human empathy with that audience ... When forged videos and images are targeted at a particular audience or utilize imagery which is bound to trigger an emotive response, deepfakes could be strategically deployed in support of inciting violence, discrediting leaders and institutions, or even tipping elections.

(Arsenault, 2020, p. 51; Chesney & Citron, 2019; Telley, 2018)

In this regard, one may mention the feasible threat of increasing availability of free deep fake services:

The so-called 'deep fakes' – digitally manipulated audio or visual material that is highly realistic and virtually indistinguishable from real material – were initially used in the movie industry. Nowadays, they are finding their application in the online realms of entertainment, consumer deception, and even politics and international affairs. Commercial and even free software are already available in the open market. It is expected that soon, the only practical constraint on one's ability to produce a deep fake will be the availability of, and access to, a sufficiently large training dataset – i.e., video and audio of the person to be modeled.

(Kertysova, 2018, p. 67; Chesney & Citron, 2019)

Additionally, scholars suggest that future technological development will reduce the costs of deepfake production and turn it more accessible, and therefore could be used by adversary actors aspiring to foster political manipulation through AI-driven deepfake (Kertysova, 2018).

Between AI technologies and the COVID-19 Pandemic

The COVID-19 pandemic has led to an initial and complex examination of AI for good. The literature has raised that AI has not been highly efficient against the pandemic. There are notable reasons mentioned, aimed at explaining this technological discrepancy, such as the needed balance between data privacy and public health and the inability to deal with changing amounts of data in different scenarios. Even though any empirical findings regarding the pandemic are still highly premature, it is suggested that the role of AI can be amplified to save lives and limit economic loss with the gathering of diagnostic data of infected individuals. Potentially, AI was estimated to being able to predict the pandemic spread over time and territory. In this regard, an AI-driven model of Health Map at Boston Children's Hospital was found more efficient in providing early notification than the Program for Monitoring Emerging Diseases (Naudé, 2020b).

Nonetheless, AI is considered less efficient for in depth tracking of COVID spread. One can explain that this is due to the lack of historical training data, and therefore at this point, there is still not enough data to build the AI models. As articulated by Naudé (2020a, p. 1):

> AI has so far not been very useful. This is for a number of reasons. The first is that AI requires data on COVID-19 to train. An example of how this can be done is the case of the 2015 Zika- virus, whose spread was ex post predicted using a dynamic neural network [...] Because COVID-19 is different from Zika, or other infections, and because there are at the time of writing still not sufficient data to build AI models that can track and forecast its spread. Most of the growing number of publications reporting on using AI for diagnostic and predictive purposes so far tend to use small, possibly biased, and mostly Chinese- based samples, and have not been peer-reviewed.

However, a successful example was raised after the spread of the Zika Virus in 2015, in which scholars were able to predict its spread using a dynamic neural network (Akhtar et al., 2019). Naudé (2020a) adds that the lack of data and the noisy social media lead to the fact that AI predictive models of COVID have not been highly reliable or accurate, and therefore most tracking models do not employ AI technologies but established epidemiological models (Wang et al., 2020). For instance, the Robert Koch Institute in Germany employs an epidemiological SIR model (Susceptible-Infected-Removed), which was also applied in China, that considers containment measures by governments, including lockdowns and quarantines (Maier & Brockmann, 2020).

Notwithstanding, there are prevalent promising AI initiatives, enabling the sharing of existing and new data and the training of new predictive models. For instance, one can mention the open access data of the GISAID Initiative, which is the former Global Initiative on Sharing All Influenza Data, and the joint initiative between the Allen Institute for Artificial Intelligence, Microsoft, Facebook and others, to establish the COVID- 9 Open Research Dataset. Notably, this dataset includes approximately 44,000 scholarly articles for data mining (Naudé, 2020a).

In addition to the above, AI can also beneficial to diagnose COVID patients. Broadly, accurate COVID diagnosis has crucial importance as it can save lives, restrict the pandemic spread and generate enough data to train AI models. An interesting example was raised in the development of the deep convolutional neural network, COVID-Net, developed by Wang, Lin, and Wong (2020). Their network has been trained on open repository data of more than 10,000 patients with various lung conditions, and therefore is capable of diagnosing COVID from chest radiography images. However, this solution still requires additional development and research, as articulated by its creators (Rawat & Wang, 2017). In addition, Yan et al. (2020) developed a prognostic prediction algorithm to predict the mortality risk of infected individuals, using machine learning. Another initiative was developed by Jiang et al. (2020), who presented an AI model that can predict whether infected people may suffer from acute respiratory distress syndrome with 80% accuracy. Important to mention, both examples were based on relatively small samples in Chinese hospitals, and therefore their predictive reliability is questionable.

Following the above, there is an additional contribution of AI to mitigate the pandemic which is reflected within the maintenance and monitoring of social control. Broadly, this aspect has some controversial interpretations as further explained by Gasser et al. (2020, p. 1):

> Data collection and processing via digital public health technologies are being promoted worldwide by governments and private companies as strategic remedies for mitigating the COVID-19 pandemic and loosening lockdown measures. However, the ethical and legal boundaries of deploying digital tools for disease surveillance and control purposes are unclear, and a rapidly evolving debate has emerged globally around the promises and risks of mobilizing digital tools for public health.

Therefore, AI-driven social control may also be correlated with mass surveillance practices. For instance, AI has been employed to scan public places for allegedly infected individuals with thermal imaging, and to enforce social distancing and lockdown measures (Rivas, 2020). In China specifically, there are infrared cameras at airports and train stations, aimed at monitoring crowds for high temperatures and they may be combined with a facial recognition system to indicate whether an individual with a high temperature had a mask (Dickson, 2020;

Chun, 2020). Another use of the system in China was to guarantee citizens obey the quarantine regulations, and based on reports, individuals who violated such restrictions, were investigated by the authorities if monitored. As raised above, such AI usage for monitoring purposes is not exclusive to China and there is a US computer vision-based startup which offers a Social Distancing Detection software. This software uses camera images to identify any alleged violations of social distancing norms (Naudé, 2020a).

Critical approach to AI governance in the public sector

Following the discussion above, policy makers and regulators aiming at reducing discriminatory patterns in society, may need to consider the tremendous impact of AI technologies on society and the ways to mitigate it. As further discussed by Hagerty and Rubinov (2019, p. 13): 'Despite high global rates of connectivity, the digital divide persists [...] The emergence of AI technologies may reinforce the current digital divide and introduce new forms of exclusion'.

Arsenault (2020) suggests that policy-makers address core liberal democratic values and take into consideration the existing social and political settings that foster disinformation and erode the democratic ethos. Among her recommendations:

> Policy responses must ensure that they do not inadvertently bolster the very narratives that they seek to disprove. For example, efforts to regulate speech, 'debunk' falsehoods, or adopt technological responses risk strengthening those narratives that seek to undermine key liberal democratic values. Policy responses must recognize that **AI-facilitated disinformation campaigns are precision-targeted**, and designed to resonate with pre-existing inclinations, biases, and beliefs; policy must therefore **address the underlying domestic contentions and fissures that can be exploited by adversarial actors**.

Nonetheless, there is an ambiguous question of whether we, as a society, aspire to create AI that truly replicates the human mind, including its defaults, or AI that reflects an enhanced version of humans (Manheim, & Kaplan, 2018). In this regard, even if one is incapable of removing bias from human beings, developers can train AI systems to be less biased with the proper governance of data samples. This notion is unique as it enables the scientific community to test crucial processes continuously under pre-determined and sterilized settings, being able to minimize the odds for human errors with machines.

In this regard, Misuraca and van Noordt (2020) posit that the scholarly discourse focuses on governance of AI, and much less on governance with AI, and describe the broad complexity revolving around the notion of AI governance in the public sector:

> [e]xplore and assess the effective use and value added of AI to redesign internal government operations and public services to better serve the citizens and businesses and enhance quality and impact of services, as well as create public private partnerships [...] better understand the potential benefits and risks of the use of AI in the public sector, and the governance mechanisms and regulatory frameworks needed to safeguard human rights and the ethical deployment of AI, especially in sensitive policy areas and domains of public interest.
>
> (p. 8)

A prevalent example of this in the European Union is shown within the Declaration on Cooperation on AI, describing the commitment to a common policy approach to facilitate the promotion of AI in the EU and address its substantial implications. Broadly, AI adoption within the public sector has tremendous potential in fostering policy making efficiency and enhanced civil services, which may increase the civil trust in the public service. In this regard, AI may be deployed in the following governmental categories, including enforcement of existing regulation and targets identification, regulatory research, analysis and monitoring of data to augment decision making processes, supporting services' delivery to the citizens and businesses and management assistance in internal organizations. Notably, Misuraca & van Noordt suggest:

> [t]he typologies of AI which are most frequently appearing in government: one related to Chatbots or Digital Assistants, and the other focused on providing some sort of intelligent, data based predictions and simulation, through the recognition and visualisation of patterns in (big) socioeconomic data [...] Most of these AI systems, 87 out of 230, are used in the provision of general public services or in communication and engagement activities. These occur in a large variety of typologies such as having Chatbots to communicate with citizens, analysing data to make public services more tailored to beneficiaries or public policy making more accessible through e.g. automatic transcription of political hearings.
>
> (2020, p. 81–82)

Broadly, as AI technologies have been applied to an abundant number of human domains, including scientific research, education, logistics, and language translation, the alleged impact of AI on societies and economies is salient at the national and international levels. Therefore, there is of paramount importance to address the notion of AI governance, in order to ensure the ethical use of AI in society, including tailored regulation and governance frameworks. ÓhÉigeartaigh et al. (2020) suggest using inter-cultural cooperation to maintain these governance frameworks. For instance, one may suggest collaboration of AI scholars globally to develop responsible AI systems, so they could share practices and experience, and potentially faster development of sophisticated AI systems. As further articulated in their study:

> Cross-cultural cooperation [...] will be essential if AI is to bring about broad benefits across societies globally, enabling advances in one part of the world to be shared with other countries, and ensuring that no part of society is neglected or disproportionately negatively impacted by AI [...] cooperation enables researchers around the world to share expertise, resources, and best practices. This enables faster progress both on beneficial AI applications, and on managing the ethical and safety-critical issues that may arise [...] in the absence of cooperation, there is a risk that competitive pressures between states or commercial ecosystems may lead to underinvestment in safe, ethical, and socially beneficial AI development [...], international cooperation is also important for more practical reasons, to ensure that applications of AI that are set to cross-national and regional boundaries (such as those used in major search engines or autonomous vehicles) can interact successfully with a sufficient range of different regulatory environments and other technologies in different regions.
>
> (p. 572)

However, society should avoid a scenario in which leading state actors impose their values on other countries (Acharya, 2019). In this regard, AI governance as a national concept should be differentiated between state actors and match the cultural values and sensitivities of the population (Hagerty & Rubinov 2019).

Notably, the adoption of AI in the public sector is complex and gradual, but the literature regarding this issue is still relatively scarce (Desouza, 2018; Williams, Brooks, & Shmargad, 2018). Different than the private sector, the public personnel and infrastructure are usually behind the technological progress, and the digitation processes are surrounded by uncertainty and skepticism. In this regard, AI has proved its ability to enhance efficiency by automating decision-making processes and services (Eggers et al., 2017). One may note prevalent case studies in the social policy, where AI predicts high risk youth (Chandler, Levitt, & List, 2011) or identifies restaurant businesses (Kang, Kuznetsova, Luca, & Choi, 2013). However, AI is also associated with job loss among the elderly community, privacy violations resulting in the facilitation of mass surveillance and the amplification of biases in policy making process (Janssen & Kuk, 2016).

As policy making processes require a stable ground of societal and regulatory norms, the rapid advancement of AI technologies and their dynamic character and the ambiguity among their definition posit some substantial challenges in the public sector. As further articulated by Krafft et al. (2020):

> Recent concern about harms of information technologies motivate consideration of regulatory action to forestall or constrain certain developments in the field of artificial intelligence (AI). However, definitional ambiguity hampers the possibility of conversation about this urgent topic of public concern. Legal and regulatory interventions require agreed-upon definitions, but consensus around a definition of AI has been elusive, especially in policy conversations [...] While these definitions are conceptually illuminating in highlighting the role of humans in AI, they are (i) too ambiguous to usefully apply in regulatory approaches to governance, and (ii) tend to misapprehend AI's current capabilities.
>
> (p. 72–73)

Another interesting finding in this study highlights the difference between the perspective of AI researchers addressing the technical functions in AI definition, compared to policy makers who emphasize the AI ability to mimic human thinking and conduct.

Among the prevalent AI obstacles for the public sphere, one can mention accuracy, bias and discrimination, legality, due process and administrative justice, responsibility, accountability, transparency and explainability, power, compliance and control. As further articulated by Henman (2020), the government uses AI technologies to improve public governance and security. In this regard, AI is used to detect real time data anomalies and cyber attacks, as well as tackling disinformation campaigns, as raised above. However, the implications of data protections, privacy violations and cyber threats are constantly increasing (Fjeld et al., 2020; Mittelstadt et al., 2016).

With the increasing role of AI-driven systems in decision-making processes and predictive models, the social and regulatory implications of AI have surfaced to vast public scrutiny. In the short term, machine learning technologies may aggravate existing discriminatory patterns and oppression against minority and disadvantaged groups, which inhibits their professional opportunities and economic growth (Buolamwini & Gebru, 2018; Dastin, 2018; Angwin, Larson, & Kirchner, 2016). Long-term and dystopian predictions suggest the option of singularity and machine-based determinism and therefore discuss the threat to human existence by delegating critical decision-making to autonomous weapons (Muller & Bostrom, 2016; Asaro, 2012). The growing public concerns regarding the detrimental AI impact have fostered a myriad of government and regulatory initiatives for AI governance (Mittelstadt, 2019).

Important to mention is that Perry and Uuk (2019) differentiate between AI technical safety and AI governance:

> AI technical safety focuses on solving computer science problems around issues like misalignment and the control problem for AGI [...] AI governance, on the other hand, studies how humanity can best navigate the transition to advanced AI systems [...] This would include the political, military, economic, governance, and ethical considerations and aspects of the problem that advanced AI has on society.
>
> (p. 3)

One of the salient obstacles for AI-driven policy making is balancing between privacy and data acquisition and fostering the proper social rules and ethical guidelines to deal with the negative implications of AI (Begg, 2008). There are notable challenges for the public sector, such as the lack of the public's trust toward AI and its ability to make decisions automatically. For instance, Sun and Medaglia (2019) highlight the importance of face-to-face interaction between patients and doctors, compared to machines. Another issue, which was highly prevalent during COVID-19 is the unethical usage and spread of personal data by third parties and malign actors (Xie, 2017).

In addition, there are additional policy and legal challenges, such as the rise of national security threats deriving from private AI companies managing huge amounts of sensitive data. As further exemplified by Sun and Medaglia (2019):

> [f]oreign company, such as IBM, collects and stores large amounts of personal data on Chinese patients. Letting a corporation of a

foreign country have access to the health records of Chinese citizens could make China more vulnerable, for instance, to biological warfare. This is considered to be no less than an existential threat for the continuation of AI in the public sector as a whole.

(p, 375)

Furthermore, one may mention the lack of AI-driven regulation decision and policy making processes. In addition, they suggest in their study to incorporate a decentralized bottom-up decision-making strategy into consolidating AI policy guidelines. Their findings showed that there is an urgent need for data integration that combines different organizations nation-wide. Notably, it is crucial to encourage the relevant stakeholders to be aware of current AI developments, in order to deal with the trust challenge.

Broadly, public AI systems are more prevalent in the health sector, whose technological investments are considered relatively high (Yang et al., 2012). An interesting study in 2019 focused on narrow AI adoption in the public sector and analysed the salient challenges of the AI system IBM Watson in public healthcare in China (Sun & Medaglia, 2019). Following the COVID-related uses, there are additional cases in which AI has shown its medical contribution (Jung & Padman, 2015). For instance, in 2010, a myriad of hospitals in England employed a disease surveillance system, using machine learning algorithms that significantly reduced the outbreaks of norovirus (Mitchell et al. 2016).

Interesting to mention is the regulatory approach suggested by Thierer, O'Sullivan and Russell (2017), highlighting the need to embrace innovation and digital experimentation by default within the policy framework for AI technologies. This approach can be tackled in case of an scenario in which technological innovation will lead to substantial harm. Meaning, AI-driven policies should be permissive towards digital innovation, and not be tempted to over restrict AI.

Following the above, we would like to reinforce their regulatory approach and suggest that the dichotomic perception of governance of AI does not allow the full contribution of human-machine teaming. In this regard, the public sector should promote governance with AI, under the core understanding that technologies can mitigate human flaws, and mutual teaming may lead to better achievements. Fostering responsible innovation governance enables society to address the core ethical and societal issues posited by emerging technologies.

References

Acharya, A. (2019). Why international ethics will survive the crisis of the liberal international order. *SAIS Review of International Affairs, 39*(1), 5–20.

Akhtar, M., Kraemer, M. U., & Goebel, L. M. (2019). A dynamic neural network model for predicting risk of Zika in real time. *BMC Medicine, 17*(1), 1–16.

Amazon Web Services. *(2017). Amazon Recognition announces real-time face recognition, Text in Image recognition, and improved face detection.* December 12, 2020, from: https://aws.amazon.com/about-aws/whats-new/2017/11/amazon-rekognition-announces-real-timeface-recognition-text-in-image-recognition-and-improved-face-detection/.

Angwin, J. Larson, J., & Kirchner, L. (2016). Machine bias: there's software used across the country to predict future criminals and it's biased against blacks. *ProPublica.* December 12, 2020, from: www.propublica.org/article/machine-bias-risk-assessments-in-criminal-sentencing.

Arsenault, A. (2020). *Microtargeting, automation, and forgery: disinformation in the age of artificial intelligence.* University of Ottawa Research. Available at: https://ruor.uottawa.ca/handle/10393/40495

Asaro. P. (2012). On banning autonomous weapon systems: human rights, automation, and the dehumanization of lethal decision making. *International Review of the Red Cross, 94*(886), 687–709.

Bartlett, R., Morse, A., Stanton, R., & Wallace, N. (2018). *Consumer-lending discrimination in the era of fintech.* Unpublished working paper. University of California.

Barocas, S., & Selbst, A. D. (2016). Big data's disparate impact. *California Law Review, 104,* 671.

Begg, R. (2008). Artificial intelligence techniques in medicine and health care. In J. Tan (Ed.). *Intelligent Information Technologies: Concepts, Methodologies, Tools, and Applications* (pp. 1750–1757). IGI Global.

Berwick, A. (2018). How ZTE helps Venezuela create China-style social control. *Reuters Investigates, 14.*

Bostrom, N. (2014). *Taking superintelligence seriously: Superintelligence: Paths, dangers, strategies.* Oxford University Press.

Buolamwini, J., & Gebru, T. (2018). Gender shades: Intersectional accuracy disparities in commercial gender classification. *Conference on Fairness, Accountability and Transparency.* Proceedings of Machine Learning Research 81, 1–15.

Cadwalladr, C. (2018). I made Steve Bannon's psychological warfare tool': meet the data war whistleblower. *Guardian.* Retrieved February 10, 2021, www.theguardian.com/news/2018/mar/17/data-war-whistleblower-christopher-wylie-faceook-nix-bannon-trump.

Chandler, D., Levitt, S. D., & List, J. A. (2011). Predicting and preventing shootings among at-risk youth. *The American Economic Review, 101*(3), 288–292.

Chesney, B., & Citron, D. (2019). Deep fakes: a looming challenge for privacy, democracy, and national security. *California Law Review, 107,* 1753.

Chun, A. (2020). In a time of coronavirus, Chinas investment in AI is paying off in a big way. *South China Morning Post, 18.*

Citron, D. K., & Pasquale, F. (2014). The scored society: due process for automated predictions. *Washington Law Review, 89,* 1–33.

Cohen, J. E. (2017). Law for the platform economy. *UCDL Review, 51,* 133.

Dastin. J. (2018). *Amazon scraps secret AI recruiting tool that showed bias against women.* Reuters. December 12, 2020, from: www.reuters.com/article/us-amazon-com-jobs-automation-insight-idUSKCN1MK08G.

Desouza, K. C. (2018). Delivering Artificial Intelligence in Government: Challenges and Opportunities. *IBM Center for The Business of Government.* December 12, 2020, from: www.businessofgovernment.org/sites/default/files/Delivering%20Artificial%20Intelligence%20in%20Government_0.pdf.

De Spiegeleire, S., Maas, M., & Sweijs, T. (2017). *Artificial intelligence and the future of defense: strategic implications for small-and medium-sized force providers.* The Hague Centre for Strategic Studies.

Dickson, B. (2020). Why AI might be the most effective weapon we have to fight COVID-19. *The Next Web, 21.*

Dougherty, C. (2015). Google photos mistakenly labels black people 'gorillas'. *The New York Times, 1.*

Eggers, W. D., Schatsky, D., & Viechnicki, P. (2017). AI-augmented government. Using cognitive technologies to redesign public sector work. Deloitte. December 12, 2020, from: https://dupress.deloitte.com/dup-us-en/focus/cognitive-technologies/artificial-intelligencegovernment.html.

Feng, E., & Lucas, L. (2018). Inside China's Surveillance State, *Financial Times,* July 20. December 12, 2020, from: www.ft.com/content/2182eebe-8a17-11e8-bf9e-8771d5404543.

Fjeld, J., Achten, N., Hilligoss, H., Nagy, A., & Srikumar, M. (2020). *Principled artificial intelligence: Mapping consensus in ethical and rights-based approaches to principles for AI.* Berkman Klein Center Research Publication No. 2020-1. https://ssrn.com/abstract=3518482

Freedom House. (2017). *The Battle for China's Spirit Religious Revival, Repression, and Resistance under Xi Jinping.* A Freedom House Special Report. December 12, 2020, from: https://freedomhouse.org/sites/default/files/FH_ChinasSprit2017_Abridged_FINAL_compressed.pdf

Friis, S. M. (2015). 'Beyond anything we have ever seen': beheading videos and the visibility of violence in the war against ISIS. *International Affairs, 91*(4), 725–746.

Gasser, U., Ienca, M., Scheibner, J., Sleigh, J., & Vayena, E. (2020). *Digital tools against COVID-19: Framing the ethical challenges and how to address them.* Cornell University. Available at: https://arxiv.org/abs/2004.10236.

Glantz, A., & Emmanuel Martinez, E. (2018). Kept out: How banks block people of color from homeownership. *Associated Press.* December 12, 2020, from: https://apnews.com/ae4b40a720b74ad8a9b0bfe65f7a9c29

Graham, D. A. (2020). Not Even Cambridge Analytica Believed Its Hype. *Atlantic.* Retrieved February 10, 2021, from: www.theatlantic.com/politics/ar-chive/2018/03/cambridge-analyticas-self-own/556016.

Greg, A., & Chan, T. (2017). Artificial Intelligence and National Security. *Belfer Center for Science and International Affairs*, Cambridge (MA), Harvard Kennedy School, 1–111.

Hagerty, A., & Rubinov, I. (2019). *Global AI ethics: a review of the social impacts and ethical implications of artificial intelligence.* Cornell University. Available at: arxiv.org/abs/1907.07892v1

Hao, K. (2018). Amazon Is the Invisible Backbone behind ICE's Immigration Crackdown. *MIT Technology Review.* December 12, 2020, from: www.technologyreview.com/s/612335/amazon-is-the-invisible-backbone-behind-ices-immigration-crackdown/.

Helmus, T. C., Bodine-Baron, E., Radin, A., Magnuson, M., Mendelsohn, J., Marcellino, W., Bega, A., & Winkelman, Z. (2018). *Russian social media influence: Understanding Russian propaganda in Eastern Europe.* Rand Corporation.

Henman, P. (2020). Improving public services using artificial intelligence: possibilities, pitfalls, governance. *Asia Pacific Journal of Public Administration, 42*(4), 209–221.

Human Rights Watch. (2018). *World Report 2018.* December 12, 2020, from: www.hrw.org/world-report/2018#

Kirkpatrick, D. (2018). 'Israeli software helped Saudi's spy on Khashoggi, lawsuit says.' *New York Times,* 9.

Kang, J. S., Kuznetsova, P., Luca, M., & Choi, Y. (2013). Where not to eat? Improving public policy by predicting hygiene inspections using online reviews. *Proceedings of the 2013 conference on empirical methods in natural language processing.* Association for Computational Linguistics (pp. 1443–1448).

Kastrenakes, J. (2018). Salesforce Employees Ask CEO to 'Re-Examine' Contract with Border Protection Agency. *The Verge.* December 12, 2020, from: www.theverge.com/2018/6/25/17504154/salesforce-employee-letter-border-protection-ice-immigration-cbp

Kay, M., Matuszek, C., & Munson, S. A. (2015, April). Unequal representation and gender stereotypes in image search results for occupations. *Proceedings of the 33rd Annual ACM Conference on Human Factors in Computing Systems.* Association for Computing Machinery (pp. 3819–3828).

Kertysova, K. (2018). Artificial Intelligence and Disinformation: How AI Changes the Way Disinformation is Produced, Disseminated, and Can Be Countered. *Security and Human Rights, 29*(1–4), 55–81.

Knight, W. (2017). The dark secret at the heart of AI. *Technology Review, 120*(3), 54–61.

Kostka, G. (2019). China's social credit systems and public opinion: Explaining high levels of approval. *New Media & Society, 21*(7), 1565–1593.

Krafft, P. M., Young, M., Katell, M., Huang, K., & Bugingo, G. (2020). Defining AI in policy versus practice. *Proceedings of the AAAI/ACM Conference on AI, Ethics, and Society.* Association for Computing Machinery (pp. 72–78).

Kreps, S. E., McCain, M., & Brundage, M. (2020). All the news that's fit to fabricate: AI-generated text as a tool of media misinformation. Cambridge University Press.

Janssen, M., & Kuk, G. (2016). The challenges and limits of big data algorithms in technocratic governance. *Government Information Quarterly, 33*(3), 371–377. https://doi.org/10.1016/j.giq.2016.08.011.

Jiang, X., Coffee, M., Bari, A., Wang, J., Jiang, X., Huang, J., Shi, J., Dai, J., Cai, J., Zhang, T., Wu, Z., He, G., & Huang, Y. (2020). Towards an artificial intelligence framework for data-driven prediction of coronavirus clinical severity. *Computers, Materials & Continua, 63*(1), 537–551.

Jung, C., & Padman, R. (2015). Disruptive digital innovation in healthcare delivery: The case for patient portals and online clinical consultations. In R. Agarwal, W. Selen, G. Roos, & R. Green (Eds.), *The Handbook of Service Innovation* (pp. 297–318). Springer London. https://doi.org/10.1007/978-1-4471-6590-3_15

Lecher, C. (2018). The employee letter denouncing Microsoft's ICE contract now has over 300 signatures. *The Verge.* December 12, 2020, from: www.theverge.com/2018/6/21/17488328/microsoft-ice-employees-signatures-protest

MacDonald, E. (2017). The fake news that sealed the fate of Antony and Cleopatra. *The Conversation,* http://theconversation.com/the-fake-news-that-sealed-the-fate-of-antony-and-cleopatra-71287.

Maier, B. F., & Brockmann, D. (2020). Effective containment explains subexponential growth in recent confirmed COVID-19 cases in China. *Science, 368*(6492), 742–746.

Manheim, K. M., & Kaplan, L. (2018). Artificial Intelligence: Risks to Privacy and Democracy. *Yale Journal of Law and Technology, 21*, 106–188.

Meserole, C., & Polyakova, A. (2019). The West is ill-prepared for the wave of 'deep fakes' that artificial intelligence could unleash. BROOKINGS. December 12, 2020, from: www.brookings.edu/blog/order-from-chaos/2018/05/25/the-west-is-ill-prepared-for-the-wave-of-deep-fakes-that-arti-ficial-intelligence-could-unleash.

Microsoft. (2018). *The Future Computed: Artificial Intelligence and its role in society.* Microsoft.

Miller, N. (2018). *Cambridge Analytica CEO Suspended After Boasts of Putting Trump in the White House.* Sydney Morning Herald. Retrieved February 10, 2021, from: www.smh.com.au/world/europe/cambridge-analytica-ceo-suspended-af-ter-boasts-of-putting-trump-in-white-house-20180321-p4z5dg.html.

Mitchell, C., Meredith, P., Richardson, M., Greengross, P., & Smith, G. B. (2016). Reducing the number and impact of outbreaks of nosocomial viral gastroenteritis: time-series analysis of a multidimensional quality improvement initiative. *BMJ Quality and Safety, 25*(6), 466–474.

Misuraca, G., & van Noordt, C. (2020). AI watch-Artificial Intelligence in public services: overview of the use and impact of AI in public services in the EU. *JRC Working Papers*, Publications Office of the European Union (JRC120399).

Mittelstadt. B. (2019). AI ethics–Too principled to fail? *Nature Machine Intelligence, 1,* 501-507.

Mittelstadt, B. D., Allo, P., Taddeo, M., Wachter, S., & Floridi, L. (2016). The ethics of algorithms: mapping the debate. *Big Data & Society, 3*(2), 2053951716679679. https://doi.org/10.1177/2053951716679679

Morris, D. Z. (2018). China will block travel for those with bad 'Social Credit'. *Fortune.* December 12, 2020, from: http://fortune.com/2018/03/18/china-travel-ban-social-credit/.

Mueller. (2019). *Report on the Investigation into Russian Interference in the 2016 Presidential Election (Vol. II).* December 12, 2020, from: www.jus-tice.gov/storage/report.pdf.

Muller, V. C., & Bostrom, N. (2016). Future progress in artificial intelligence: A survey of expert opinion. In V. Müller (Ed.), Fundamental Issues of Artificial Intelligence. Synthese Library (Studies in Epistemology, Logic, Methodology, and Philosophy of Science), vol. 376. Springer. 555–572.

Perry, B., & Uuk, R. (2019). AI governance and the policymaking process: key considerations for reducing AI risk. *Big Data and Cognitive Computing, 3*(2), 26.

Nafeez, A. (2018). Pentagon Wants to Predict Anti-Trump Protests Using Social Media Surveillance. *Motherboard.* December 12, 2020, from: https://motherboard.vice.com/en_us/article/7x3g4x/pentagon-wants-to-predict-anti-trump-protestsusing-social-media-surveillance.

Naudé, W. (2020a). Artificial Intelligence against COVID-19: An early review. IZA Discussion Paper no. 13110, Bonn.

Naudé, W. (2020b). Artificial intelligence vs COVID-19: limitations, constraints and pitfalls. *AI & Society, 35*(3), 761–765.

Nebehay, S. (2018). UN calls for Myanmar generals to be tried for genocide, blames Facebook for incitement. *Reuters.* 27 August.

Ochigame, R., & Holston, J. (2016). Filtering Dissent Social Media and Land Struggles in Brazil. *New Left Review, 99,* 85–110.

ÓhÉigeartaigh, S. S., Whittlestone, J., Liu, Y., Zeng, Y., & Liu, Z. (2020). Overcoming barriers to cross-cultural cooperation in AI ethics and governance. *Philosophy & Technology, 33*(4), 571–593.

Pasquale, F. (2015). *The Black Box Society: The Secret Algorithms That Control Money and Information.* Harvard University Press.

Patterson, D. (2018). *How AI is Creating New Threats to Election Security,* CBS NEWs. December 12, 2020, from: www.cbsnews.com/news/how-ai-will-shape-the-future-of-election-security.

Polonski, V. (2017, August). How artificial intelligence silently took over democracy. In *World Economic Forum.* December 12, 2020, from: www.wefo-rum.org/agenda/2017/08/artificial-intelligence-can-save-democracy-unless-it-destroys-it-first

Ram, N. (2017). Innovating Criminal Justice. *Northwestern University Law Review, 112*(4), 659–724.

Rawat, W., & Wang, Z. (2017). Deep convolutional neural networks for image classification: A comprehensive review. *Neural Computation, 29*(9), 2352–2449.

Rivas, A. (2020). Drones and artificial intelligence to enforce social isolation during COVID-19 outbreak. *Medium Towards Data Science, 26.*

Roose, K. (2018). Facebook Grapples With a Maturing Adversary in Election Meddling. *New York Times, 1.*

Samuel, S. (2018). China Is Going to Outrageous Lengths to Surveil Its Own Citizens. *The Atlantic.* December 12, 2020, from: www.theatlantic.com/international/archive/2018/08/china-surveillance-technology-muslims/567443/

Schiff, D., Biddle, J., Borenstein, J., & Laas, K. (2020, February). What's next for AI ethics, policy, and governance? A global overview. *Proceedings of the AAAI/ACM Conference on AI, Ethics, and Society.* Association for Computational Linguistics (pp. 153–158).

Shaban, H. (2018). Amazon employees demand company cut ties with ICE. *Washington Post, 22.*

Shen, A. (2018). Facial Recognition Tech Comes to Hong Kong-Shenzhen Border. *South China Morning Post.* December 12, 2020, from: www.scmp.com/news/china/society/article/2156510/china-uses-facial-recognition-system-deter-tax-free-traders-hong.

Sonnad, N. (2018). A Flawed Algorithm Led the UK to Deport Thousands of Students. *Quartz.* December 12, 2020, from: https://qz.com/1268231/a-toeic-test-led-the-uk-to-deport-thousands-of-students/

State Council Notice Concerning Issuance of the Planning Outline for the Establishment of a Social Credit System (2014–2020). December 12, 2020, from: https://chinacopyrightandmedia.wordpress.com/2014/06/14/planning-outline-for-the-construction-of-a-social-credit-system-2014-2020/

Stecklow, S. (2018), Why Facebook is losing the war on hate speech in Myanmar. *Reuters.* December 12, 2020, from: www.reuters.com/investigates/special-report/myanmar-facebook-hate/

Sun, T. Q., & Medaglia, R. (2019). Mapping the challenges of Artificial Intelligence in the public sector: Evidence from public healthcare. *Government Information Quarterly, 36*(2), 368–383.

Tazhiyeva, A. (2018). Challenges and opportunities of introducing Internet of Things and Artificial Intelligence applications into Supply Chain Management. *University of Vaasa.* Available at: https://osuva.uwasa.fi/handle/10024/5810

Telley, C. (2018). The Influence Machine: Automated Information Operations as a Strategic Defeat Mechanism. *National Security Affairs: The Land Warfare Papers, 121,* 1–11.

Thierer, A. D., Castillo O'Sullivan, A., & Russell, R. (2017). Artificial intelligence and public policy. *Mercatus Research Paper.*

Timberg, C., & Harris, S. (2018). Russian operatives blasted 18,000 tweets ahead of a huge news day during the 2016 presidential campaign. Did they know what was coming. *Washington Post.* December 12, 2020, from: www.washingtonpost.com/technol-ogy/2018/07/20/russian-operatives-blasted-tweets-ahead-huge-news-day-during-presidential-campaign-did-they-know-what-was-coming.

Vanderklippe, N. (2018). Chinese blacklist an early glimpse of sweeping new social- credit control. *The Globe and Mail, 3.*

Waddell, K. (2016). How algorithms can bring down minorities' credit scores. *The Atlantic*, 2.

Wang, L., Lin, Z. Q., & Wong, A. (2020). Covid-net: A tailored deep convolutional neural network design for detection of covid-19 cases from chest x-ray images. *Scientific Reports*, *10*(1), 1–12.

Wang, L., Zhou, Y., He, J., Zhu, B., Wang, F., Tang, L., Eisenberg, M., & Song, P. X. (2020). An epidemiological forecast model and software assessing interventions on the COVID-19 epidemic in China. *Journal of Data Science*, *18*(3), 409–432.

Weiner, J. (2018). ACLU: Amazon's Face-Recognition Software Matched Members of Congress with Mugshots. *Orlando Sentinel*. December 12, 2020, from: www.orlandosentinel.com/news/politics/political-pulse/os-amazon-rekognition-face-matching-software-congress-20180726-story.html

Whittaker, M., Crawford, K., Dobbe, R., Fried, G., Kaziunas, E., Mathur, V., West, S. M., Richardson, R., Schultz, J., & Schwartz, O. (2018). *AI Now Report 2018*. AI Now Institute, New York University.

Williams, B. A., Brooks, C. F. & Shmargad, Y. (2018). How algorithms discriminate based on data they lack: Challenges, solutions, and policy implications. *Journal of Information Policy, 8*, 78–115;

World Economic Forum. (2018). How to Prevent Discriminatory Outcomes in Machine Learning. December 12, 2020, from: www3.weforum.org/docs/WEF_40065_White_Paper_How_to_Prevent_Discriminatory_Outcomes_in_Machine_Learning.pdf.

Xie, G. (2017). Rèn zhī yīliáo: Gèxìng huà, xún zhèng de zhìhuì yīliáo (CognitiveCare: Personalized, evidence-based smart medicine). Presented at the AI public courses in Peking University. December 12, 2020, from: https://mp.weixin.qq.com/s/G8p_js1HXK03Us7jbNUNUg

Yang, Z., Ng, B.-Y., Kankanhalli, A., & Luen Yip, J. W. (2012). Workarounds in the use of IS in healthcare: A case study of an electronic medication administration system. *International Journal of Human-Computer Studies*, *70*(1), 43–65. https://doi.org/10.1016/j.ijhcs.2011.08.002

Yan, L., Zhang, H. T., Xiao, Y., Wang, M., Guo, Y., Sun, C., … & Yuan, Y. (2020). Prediction of criticality in patients with severe Covid-19 infection using three clinical features: a machine learning-based prognostic model with clinical data in Wuhan. *Available at:* www.medrxiv.org/content/10.1101/2020.02.27.20028027v2

Index